Yearbook of Astronomy 2025

Front Cover: Seen here in an image captured by the James Webb Space Telescope's Mid-Infrared Instrument and located at a distance of around 40 million light-years in the constellation of Doradus, NGC 1566 is a very dusty spiral galaxy, with an extremely bright nucleus typical of the Seyfert class of galaxies. The reddish areas show where star formation is occurring. (NASA/ESA/CSA/Judy Schmidt)

YEARBOOK OF
ASTRONOMY
2025

EDITED BY

Brian Jones

WHITE OWL

AN IMPRINT OF PEN & SWORD BOOKS LTD.
YORKSHIRE – PHILADELPHIA

First published in Great Britain in 2024 by
WHITE OWL
An imprint of
Pen & Sword Books Ltd
Yorkshire – Philadelphia

ISBN 978 1 03611 515 9

Typeset in Dante By Mac Style
Printed and bound in India by Parksons Graphics Pvt. Ltd.

Pen & Sword Books Ltd. incorporates the imprints of Pen & Sword Books:
After the Battle, Archaeology, Atlas, Aviation, Battleground, Discovery,
Family History, History, Maritime, Military, Politics, Select, Transport,
True Crime, Fiction, Frontline Books, Leo Cooper, Praetorian Press,
Seaforth Publishing, Wharncliffe and White Owl.

For a complete list of Pen & Sword titles please contact

PEN & SWORD BOOKS LIMITED
George House, Beevor Street, Off Pontefract Road, Hoyle Mill,
Barnsley, South Yorkshire, England, S71 1HN.
E-mail: enquiries@pen-and-sword.co.uk
Website: www.pen-and-sword.co.uk

or

PEN AND SWORD BOOKS
1950 Lawrence Rd, Havertown, PA 19083, USA
E-mail: Uspen-and-sword@casematepublishers.com
Website: www.penandswordbooks.com

Contents

Article Section

Miscellaneous

Editor's Foreword

The *Yearbook of Astronomy 2025* is the latest edition of what has long been an indispensable publication, the annual appearance of which has been eagerly anticipated by astronomers, both amateur and professional, for well over half a century. As ever, the *Yearbook* is aimed at both the armchair astronomer and the active backyard observer. Within its pages you will find a rich blend of information, star charts and guides to the night sky coupled with an interesting mixture of articles which collectively embrace a wide range of topics, ranging from the history of astronomy to the latest results of astronomical research; space exploration to observational astronomy; and our own celestial neighbourhood out to the farthest reaches of space.

The *Monthly Star Charts* have been compiled by David Harper and show the night sky as seen throughout the year. Two sets of twelve charts have been provided, one set for observers in the Northern Hemisphere and one for those in the Southern Hemisphere. Between them, each pair of charts depicts the entire sky as two semi-circular half-sky views, one looking north and the other looking south.

A new feature for the Yearbook of Astronomy is a set of pages called *Lunar Phenomena*, compiled by Lynne Marie Stockman, which includes the phases and occultations (and eclipses) as in previous editions but which now contains apsides and node crossings with references to eclipses. There is also an item on lunistices, or lunar standstills, due to there being a major lunar standstill in 2025. The next similar event, a minor lunar standstill, will occur in 2034, so this particular section will be omitted from the *Lunar Phenomena* pages until then. This section of the Yearbook concludes with details of *Eclipses in 2025*.

This is followed by general summaries of the observing conditions for each of the planets in *The Planets in 2025*. Apparition charts for all the major planetary members of our Solar System have been compiled by David Harper. Further details of these are given in the article *Using the Yearbook of Astronomy as an Observing Guide*. The finder charts for the superior planets follow the article *The Planets in 2025*, those depicting the morning and evening apparitions of Mercury and Venus being scattered throughout the *Monthly Sky Notes*.

As with *The Planets in 2025*, the *Monthly Sky Notes* have been compiled by Lynne Marie Stockman and give details of the positions and visibility of the planets for each month throughout 2025. At the beginning of each of the monthly notes is a list of the significant Solar System events occurring during that particular month, and which collectively replace the single list of events that has been a feature in previous editions of the *Yearbook of Astronomy*. Each section of the *Monthly Sky Notes* is accompanied by a short article, the range of which includes items on a variety of astronomy- and space-related topics including an interesting item by Lynne, in which we discover that, although constellations come and constellations go, some leave behind traces of their former existence. In the second instalment of her *Gone But Not Forgotten* series, Lynne examines the tiny but interesting constellation Anser (the Goose), originally created by Polish astronomer Johannes Hevelius as an offering from Vulpecula (the Little Fox, another Hevelius creation) to Cerberus (the three-headed dog of Hades, yet another Hevelius creation) who is about to be despatched by Hercules.

The *Monthly Sky Notes and Articles* section of the book concludes with a trio of articles penned by Neil Norman, these being *Comets in 2025*, *Minor Planets in 2025* and *Meteor Showers in 2025*, all three titles being fairly self-explanatory describing as they do the occurrence and visibility of examples of these three classes of object during and throughout the year.

In his article *Recent Advances in Astronomy*, regular contributor Rod Hine looks back at the very successful first year of science from the James Webb Space Telescope, particularly the way the new views in infra-red reveal more details about objects first observed long ago. Some new ideas about the formation of planets may challenge the previously accepted timescales, and yet another unexpected result with "Dark Matter", or rather the lack of, has caused much controversy. The venerable HST still brings up interesting observations and the first tentative observations of very long period gravitational waves have been published.

This is followed by *Recent Advances in Solar System Exploration* in which Peter Rea updates us on the progress of a number of planetary missions. Mars continues to be a favourite target for exploration, the next major mission to Mars being a joint NASA/ESA venture to return to Earth samples of Martian soil and rocks collected by the Perseverance rover. In the meantime, Mercury awaits the arrival of the ESA/JAXA BepiColombo mission in 2025, while Jupiter's icy moons will be a target for ESA's JUICE and NASA's Europa Clipper spacecraft. In addition to all this, the period from 2020 to 2030 could well be called the "Decade of the Moon", with many unmanned robotic missions being sent to the lunar surface. Under the Commercial Lunar Payload Services program a number of NASA sponsored missions will be sent to the Moon in support of the Artemis program, which will return humans to the surface of the Moon by the mid 2020s.

Why do we name stars with Greek letters? In his article *Anniversaries in 2025* Neil Haggath tells us about the man who established the system, Johann Bayer, who died 400 years ago. Also commemorated are the founding of the Royal Observatory in 1675, the centenary of the death of French astronomy populariser Camille

Flammarion, and the first joint American-Russian space mission, the Apollo-Soyuz Test Project, 50 years ago.

In *Skies Over Ancient America: Mystical Mounds and Landmarks of the Prehistoric Americas* – the third and final part of his series of articles relating to Native American astronomy – P. Clay Sherrod tells us about what appear to be the earliest origins by the ancient cultures of the western hemisphere to both preserve the mythological and religious significance of the celestial realm, as well as development of elaborate structures, buildings, temples and calendars to monitor and predict the passage of time. After the sudden advent of agriculture in what had been a primitive hunter-gatherer environment, and as the climate warmed into the northern latitudes of the western hemisphere, the need for accurate prediction of our concept of TIME became a reality. This final segment examines some of the most elaborate – as well as the quite simple – beginnings of timekeeping and sky worship in the Western Hemisphere.

In his article *Astrophysicist Cecilia Helena Payne and Professor H. N. Russell*, regular contributor David M. Harland looks back at the pioneering research of the young English astronomer Cecilia Payne at Harvard University a century ago, indicating that stars are made principally of hydrogen. Her finding – which was initially rejected by the leading astrophysicists of the time in both America and Britain, but was ultimately proved correct – led a later giant in the field to declare her work as being, "... undoubtedly the most brilliant Ph.D. thesis ever written in astronomy."

It is often said that 'two heads are better than one' and the same could be said of stars. Binaries offer astronomers a unique opportunity to measure the stars' vital statistics and calibrate stellar evolution models. In her article *The Astronomers' Stars: The Terrible Twos* Lynne Marie Stockman takes a look at some of the more unusual members of this class. From the enormous binaries named after Pearce and Plaskett to the binary pulsar that led to a Nobel Prize, from the binary white dwarf system with a trick up its sleeve to the enigmatic SS 433, we explore these amazing stars named after their discoverers.

This is followed by *Mission to Mars: Countdown to Building a Brave New World: It's Life, But Not as We Know It* by Martin Braddock, the fifth in a series of articles scheduled to appear in the *Yearbook of Astronomy* throughout the 2020s and which will keep the reader fully up to date with the ongoing preparations geared towards sending a manned mission to Mars at or around the turn of the decade.

Next we have *Eta Carinae: Chance Encounter and Journey of Discovery* in which Peter Rea tells us of a lucky photo he took of the New Zealand night sky in 2004 which happened to show a pinkish fuzzy patch. This turned out to be the Eta Carinae Nebula, unknown to Peter at the time. On his journey of discovery Peter tells us about this amazing nebula. Also discussed are the observations and drawings of Eta Carinae made by Sir John Herschel from his Feldhausen estate in Cape Colony, now the Republic of South Africa, where he also observed "The Great Eruption" within Eta Carinae, remnants of which can still be seen today as the Homunculus Nebula.

This year, the Earth and the Sun cross Saturn's ring-plane, and the rings briefly become (almost) invisible. Famous astronomers including Jean-Dominique Cassini, William Herschel, Edward Emerson Barnard and Audouin Dollfus have observed ring-plane crossings over the past four centuries, often making important discoveries. David Harper explores this fascinating story in his article *Saturn at its Equinox: A History of Ring-Plane Crossings from 1612 to 2025*. In *Saturn at its Equinox: The Ring-Plane Crossing of 2025*, following the May sky notes, he explains what observers of Saturn can expect to see during this year's journey through the ring plane.

The article *A History of Observatory Designs: Before the Telescope* by Katrin Raynor is the first of a series relating to historic and innovative observatory designs. Our night sky has been a cause for wonder for thousands of years. Prehistoric archaeological sites of ancient observatories constructed of carefully placed stones are evidence that our ancestors were looking to the sky and recording their observations. Thousands of years later, observatories were designed by astronomers and built by skilled architects. This article takes the reader on a journey from Stonehenge in Wiltshire to the last significant observatory built in Europe prior to the invention of the telescope.

In *Signals from the Magnetosphere* John Vetterlein introduces us to the beauty and intricacies of the aurorae, the article being primarily based on observations carried out from his home on the island of Rousay in the Orkney Islands of Scotland. The imagery that he has captured over the years clearly illustrates the beauty of this wonderful atmospheric phenomenon that has fascinated and beguiled people for thousands of years, and continues to do so.

In the article *How to Read a Scientific Paper*, David Harper explains how professional astronomers communicate their research through journals such as *Monthly Notices of the Royal Astronomical Society*. The Internet has made this rich source of information much more accessible to amateur astronomers, and David offers practical advice on how to make good use of it.

Penned by regular contributor John McCue, the article *Small Stars* describes the smallest stars known to astronomers – neutron stars. We learn of their structure and form, their disguises, how we observe them, and of the people who set the scene for their discovery. The article also attempts to put neutron stars in the context of our daily lives, even though they are dramatic in the extreme.

The final section of the book starts off with *Some Interesting Variable Stars* by Tracie Heywood which contains useful information on variables as well as predictions for timings of minimum brightness of the famous eclipsing binary Algol for 2025. *Some Interesting Double Stars* and *Some Interesting Nebulae, Star Clusters and Galaxies* present a selection of objects for you to seek out in the night sky. The lists included here are by no means definitive and may well omit your favourite celestial targets. If this is the case, please let us know and we will endeavour to include these in future editions of the *Yearbook of Astronomy*.

Next we have a selection of *Astronomical Organizations*, which lists organizations and associations across the world through which you can further pursue your interest and participation in astronomy (if there are any that we have omitted please let us know) and *Our Contributors*, which contains brief background details of the numerous writers who have contributed to this edition of the *Yearbook*.

New topics and themes are occasionally introduced into the *Yearbook of Astronomy*, allowing it to keep pace with the increasing range of skills, techniques and observing methods now open to amateur astronomers, this in addition to articles relating to our rapidly-expanding knowledge of the Universe in which we live. There is always an interesting mix, some articles written at a level which will appeal to the casual reader and some of what may be loosely described as at a more academic level. The intention is to fully maintain and continually increase the usefulness and relevance of the *Yearbook of Astronomy* to the interests of the readership who are, without doubt, the most important aspect of the *Yearbook* and the reason it exists in the first place. With this in mind, suggestions from readers for further improvements and additions to the *Yearbook* content are welcomed. All thoughts and comments can be sent via the *Yearbook of Astronomy* website at **yearbookofastronomy.com** After all, the book is written for you ...

As ever, grateful thanks are extended to those individuals who have contributed a great deal of time and effort to the *Yearbook of Astronomy 2025*, including David Harper, who has provided updated versions of his *Monthly Star Charts*. These were generated specifically for what has been described as the new generation of the *Yearbook of Astronomy*, and the charts add greatly to the overall value of the book to star gazers. Equally important are the efforts of Lynne Marie Stockman who has put together the *Monthly Sky Notes*. Their combined efforts have produced what can justifiably be described as the backbone of the *Yearbook of Astronomy*. Also worthy of mention is Mat Blurton, who has done an excellent job typesetting the *Yearbook*, and David M. Harland who has provided valuable assistance with several images used in this edition. Thanks are also to Nick Bull of Stonehenge Dronescapes Photography for supplying the images of Stonehenge appearing in the article *A History of Observatory Designs* and Guy Wells of Northolt Branch Observatories for two of the images featuring in *Minor Planets in 2025*. Also Jonathan Wright, Charlotte Mitchell, Lori Jones, Janet Brookes, Paul Wilkinson, Charlie Simpson and Rosie Crofts of Pen & Sword Books Ltd for their efforts in producing and promoting the *Yearbook of Astronomy 2025*, the latest edition of this much-loved and iconic publication.

Brian Jones - Editor
Bradford, West Riding of Yorkshire
November 2023

Preface

The information given in this edition of the *Yearbook of Astronomy* is in narrative form. The positions of the planets given in the *Monthly Sky Notes* often refer to the constellations in which they lie at the time. These can be found on the star charts which collectively show the whole sky via two charts depicting the northern and southern circumpolar stars and forty-eight charts depicting the main stars and constellations for each month of the year. The northern and southern circumpolar charts show the stars that are within 45° of the two celestial poles, while the monthly charts depict the stars and constellations that are visible throughout the year from Europe and North America or from Australia and New Zealand. The monthly charts overlap the circumpolar charts. Wherever you are on the Earth, you will be able to locate and identify the stars depicted on the appropriate areas of the chart(s).

There are numerous star atlases available that offer more detailed information, such as *Sky & Telescope's POCKET SKY ATLAS* and *Norton's STAR ATLAS and Reference Handbook* to name but a couple. In addition, more precise information relating to planetary positions and so on can be found in a number of publications, a good example of which is *The Handbook of the British Astronomical Association*, as well as many of the popular astronomy magazines such as the British monthly periodicals *Sky at Night* and *Astronomy Now* and the American monthly magazines *Astronomy* and *Sky & Telescope*.

About Time

Before the late eighteenth century, the biggest problem affecting mariners sailing the seas was finding their position. Latitude was easily determined by observing the altitude of the pole star above the northern horizon. Longitude, however, was far more difficult to measure. The inability of mariners to determine their longitude often led to them getting lost, and on many occasions shipwrecked. To address this problem King Charles II established the Royal Observatory at Greenwich in 1675 and from here, Astronomers Royal began the process of measuring and cataloguing the stars as they passed due south across the Greenwich meridian.

Now mariners only needed an accurate timepiece (the chronometer invented by Yorkshire-born clockmaker John Harrison) to display GMT (Greenwich Mean Time). Working out the local standard time onboard ship and subtracting this from GMT gave the ship's longitude (west or east) from the Greenwich meridian. Therefore mariners always knew where they were at sea and the longitude problem was solved.

Astronomers use a time scale called Universal Time (UT). This is equivalent to Greenwich Mean Time and is defined by the rotation of the Earth. The *Yearbook of Astronomy* gives all times in UT rather than in the local time for a particular city or country. Times are expressed using the 24-hour clock, with the day beginning at midnight, denoted by 00:00. Universal Time (UT) is related to local mean time by the formula:

Local Mean Time = UT – west longitude

In practice, small differences in longitude are ignored and the observer will use local clock time which will be the appropriate Standard (or Zone) Time. As the formula indicates, places in west longitude will have a Standard Time slow on UT, while those in east longitude will have a Standard Time fast on UT. As examples we have:

Standard Time in

New Zealand	UT +12 hours
Victoria, NSW	UT +10 hours
Japan	UT + 9 hours
Western Australia	UT + 8 hours
India	UT + 5 hours 30 minutes
Pakistan	UT + 5 hours
Kenya	UT + 3 hours
South Africa	UT + 2 hours
British Isles	UT
Newfoundland Standard Time	UT − 3 hours 30 minutes
Atlantic Standard Time	UT − 4 hours
Eastern Standard Time	UT − 5 hours
Central Standard Time	UT − 6 hours
Mountain Standard Time	UT − 7 hours
Pacific Standard Time	UT − 8 hours
Alaska Standard Time	UT − 9 hours
Hawaii-Aleutian Standard Time	UT − 10 hours

During the periods when Summer Time (also called Daylight Saving Time) is in use, one hour must be added to Standard Time to obtain the appropriate Summer/ Daylight Saving Time. For example, Pacific Daylight Time is UT − 7 hours.

Using the Yearbook of Astronomy as an Observing Guide

Notes on the Monthly Star Charts

The star charts on the following pages show the night sky throughout the year. There are two sets of charts, one for use by observers in the Northern Hemisphere and one for those in the Southern Hemisphere. The first set is drawn for latitude 52°N and can be used by observers in Europe, Canada and most of the United States. The second set is drawn for latitude 35°S and show the stars as seen from Australia and New Zealand. Twelve pairs of charts are provided for each of these latitudes.

Each pair of charts shows the entire sky as two semi-circular half-sky views, one looking north and the other looking south. A given pair of charts can be used at different times of year. For example, chart 1 shows the night sky at midnight on 21 December, but also at 2am on 21 January, 4am on 21 February and so forth. The accompanying table will enable you to select the correct chart for a given month and time of night. The caption next to each chart also lists the dates and times of night for which it is valid.

The charts are intended to help you find the more prominent constellations and other objects of interest mentioned in the monthly observing notes. To avoid the charts becoming too crowded, only stars of magnitude 4.5 or brighter are shown. This corresponds to stars that are bright enough to be seen from any dark suburban garden on a night when the Moon is not too close to full phase.

Each constellation is depicted by joining selected stars with lines to form a pattern. There is no official standard for these patterns, so you may occasionally find different patterns used in other popular astronomy books for some of the constellations.

Any map projection from a sphere onto a flat page will by necessity contain some distortions. This is true of star charts as well as maps of the Earth. The distortion on the half-sky charts is greatest near the semi-circular boundary of each chart, where it may appear to stretch constellation patterns out of shape.

The charts also show selected deep-sky objects such as galaxies, nebulae and star clusters. Many of these objects are too faint to be seen with the naked eye, and you will need binoculars or a telescope to observe them. Please refer to the table of deep-sky objects for more information.

Planetary Apparition Diagrams

The diagrams of the apparitions of Mercury and Venus show the position of the respective planet in the sky at the moment of sunrise or sunset throughout the entire apparition. Two sets of positions are plotted on each chart: for latitude 52° North (blue line) and for latitude 35° South (red line). A thin dotted line denotes the portion of the apparition which falls outside the year covered by this edition of the *Yearbook*. A white dot indicates the position of Venus on the first day of each month, or of Mercury on the first, eleventh and 21st of the month. The day of greatest elongation (GE) is also marked by a white dot. Note that the dots do NOT indicate the magnitude of the planet.

Mars begins the year in retrograde in Cancer, crossing into Gemini on 12 January four days before opposition on the 16th. It resumes direct motion on 24 February. The first finder chart for Mars shows the entire retrograde loop, from 1 October 2024 to 1 June 2025. For most of this year, Mars is an evening object, moving through Gemini, Cancer, Leo and Virgo, until it is lost in the evening twilight in October. The second finder chart shows the path of Mars through these constellations from 1 January to 1 October. The ecliptic is denoted by the dotted line running across the centre of the chart from left to right. In both charts, the position of Mars is indicated on the 1st of each month as well as at opposition and at stationary points in Right Ascension. Note that the dots do NOT indicate the magnitude of Mars.

The finder charts for Jupiter, Saturn, Uranus and Neptune show the paths of the planets throughout the year. The position of each planet is indicated at opposition and at stationary points, as well as the start and end of the year and on the 1st of each month (1st of April, July and October only for Uranus and Neptune) where these dates do not fall too close to an event that is already marked. Stars are shown to magnitude 5.5 on the charts for Jupiter and Saturn. On the Uranus chart, stars are shown to magnitude 8; on the Neptune chart, the limiting magnitude is 10. In both cases, this is approximately two magnitudes fainter than the planet itself. Right Ascension and Declination scales are shown for the epoch J2000 to allow comparison with modern star charts. Note that the sizes of the dots denoting the planets do NOT indicate their magnitudes.

Selecting the Correct Charts

The table below shows which of the charts to use for particular dates and times throughout the year and will help you to select the correct pair of half-sky charts for any combination of month and time of night.

The Earth takes 23 hours 56 minutes (and 4 seconds) to rotate once around its axis with respect to the fixed stars. Because this is around four minutes shorter than a full 24 hours, the stars appear to rise and set about 4 minutes earlier on each successive day, or around an hour earlier each fortnight. Therefore, as well as showing the stars at 10pm (22h in 24-hour notation) on 21 January, chart 1 also depicts the sky at 9pm (21h) on 6 February, 8pm (20h) on 21 February and 7pm (19h) on 6 March.

The times listed do not include summer time (daylight saving time), so if summer time is in force you must subtract one hour to obtain standard time (GMT if you are in the United Kingdom) before referring to the chart. For example, to find the correct chart for mid-September in the northern hemisphere at 3am summer time, first of all subtract one hour to obtain 2am (2h) standard time. Then you can consult the table, where you will find that you should use chart 11.

The table does not indicate sunrise, sunset or twilight. In northern temperate latitudes, the sky is still light at 18h and 6h from April to September, and still light at 20h and 4h from May to August. In Australia and New Zealand, the sky is still light at 18h and 6h from October to March, and in twilight (with only bright stars visible) at 20h and 04h from November to January.

Local Time	18h	20h	22h	0h	2h	4h	6h
January	11	12	1	2	3	4	5
February	12	1	2	3	4	5	6
March	1	2	3	4	5	6	7
April	2	3	4	5	6	7	8
May	3	4	5	6	7	8	9
June	4	5	6	7	8	9	10
July	5	6	7	8	9	10	11
August	6	7	8	9	10	11	12
September	7	8	9	10	11	12	1
October	8	9	10	11	12	1	2
November	9	10	11	12	1	2	3
December	10	11	12	1	2	3	4

Legend to the Star Charts

STARS		DEEP-SKY OBJECTS	
Symbol	Magnitude	Symbol	Type of object
•	0 or brighter	※	Open star cluster
•	1	◌	Globular star cluster
•	2	□	Nebula
•	3	▦	Cluster with nebula
·	4	○	Planetary nebula
·	5	◌	Galaxy
✦	Double star		Magellanic Clouds
◉	Variable star		

Star Names

There are over 200 stars with proper names, most of which are of Roman, Greek or Arabic origin although only a couple of dozen or so of these names are used regularly. Examples include Arcturus in Boötes, Castor and Pollux in Gemini and Rigel in Orion.

A system whereby Greek letters were assigned to stars was introduced by the German astronomer and celestial cartographer Johann Bayer in his star atlas *Uranometria*, published in 1603. Bayer's system is applied to the brighter stars within any particular constellation, which are given a letter from the Greek alphabet followed by the genitive case of the constellation in which the star is located. This genitive case is simply the Latin form meaning 'of' the constellation. Examples are the stars Alpha Boötis and Beta Centauri which translate literally as 'Alpha of Boötes' and 'Beta of the Centaur'.

As a general rule, the brightest star in a constellation is labelled Alpha (α), the second brightest Beta (β), and the third brightest Gamma (γ) and so on, although there are some constellations where the system falls down. An example is Gemini where the principal star (Pollux) is designated Beta Geminorum, the second brightest (Castor) being known as Alpha Geminorum.

There are only 24 letters in the Greek alphabet, the consequence of which was that the fainter naked eye stars needed an alternative system of classification. The system in popular use is that devised by the first Astronomer Royal John Flamsteed in which the stars in each constellation are listed numerically in order from west to east. Although many of the brighter stars within any particular constellation will have both Greek letters and Flamsteed numbers, the latter are generally used only when a star does not have a Greek letter.

The Greek Alphabet

α	Alpha	ι	Iota	ρ	Rho
β	Beta	κ	Kappa	σ	Sigma
γ	Gamma	λ	Lambda	τ	Tau
δ	Delta	μ	Mu	υ	Upsilon
ε	Epsilon	ν	Nu	φ	Phi
ζ	Zeta	ξ	Xi	χ	Chi
η	Eta	o	Omicron	ψ	Psi
θ	Theta	π	Pi	ω	Omega

The Names of the Constellations

On clear, dark, moonless nights, the sky seems to teem with stars although in reality you can never see more than a couple of thousand or so at any one time when looking with the unaided eye. Each and every one of these stars belongs to a particular constellation, although the constellations that we see in the sky, and which grace the pages of star atlases, are nothing more than chance alignments. The stars that make up the constellations are often situated at vastly differing distances from us and only appear close to each other, and form the patterns that we see, because they lie in more or less the same direction as each other as seen from Earth.

A large number of the constellations are named after mythological characters, and were given their names thousands of years ago. However, those star groups lying close to the south celestial pole were discovered by Europeans only during the last few centuries, many of these by explorers and astronomers who mapped the stars during their journeys to lands under southern skies. This resulted in many of the newer constellations having modern-sounding names, such as Octans (the Octant) and Microscopium (the Microscope), both of which were devised by the French astronomer Nicolas Louis De La Caille during the early 1750s.

Over the centuries, many different suggestions for new constellations have been put forward by astronomers who, for one reason or another, felt the need to add new groupings to star charts and to fill gaps between the traditional constellations. Astronomers drew up their own charts of the sky, incorporating their new groups into them. A number of these new constellations had cumbersome names, notable examples including Officina Typographica (the Printing Shop) introduced by the German astronomer Johann Bode in 1801; Sceptrum Brandenburgicum (the Sceptre of Brandenburg) introduced by the German astronomer Gottfried Kirch in 1688; Taurus Poniatovii (Poniatowski's Bull) introduced by the Polish-Lithuanian astronomer Martin Odlanicky Poczobut in 1777; and Quadrans Muralis (the Mural Quadrant) devised by the French astronomer Joseph-Jerôme de Lalande in 1795. Although these have long since been rejected, the latter has been immortalised by the annual Quadrantid meteor shower, the radiant of which lies in an area of sky formerly occupied by Quadrans Muralis.

During the 1920s the International Astronomical Union (IAU) systemised matters by adopting an official list of 88 accepted constellations, each with official spellings and abbreviations. Precise boundaries for each constellation were then drawn up so that every point in the sky belonged to a particular constellation.

The abbreviations devised by the IAU each have three letters which in the majority of cases are the first three letters of the constellation name, such as AND for Andromeda, EQU for Equuleus, HER for Hercules, ORI for Orion and so on. This trend is not strictly adhered to in cases where confusion may arise. This happens with the two constellations Leo (abbreviated LEO) and Leo Minor (abbreviated LMI). Similarly, because Triangulum (TRI) may be mistaken for Triangulum Australe, the latter is abbreviated TRA. Other instances occur with Sagitta (SGE) and Sagittarius (SGR) and with Canis Major (CMA) and Canis Minor (CMI) where the first two letters from the second names of the constellations are used. This is also the case with Corona Australis (CRA) and Corona Borealis (CRB) where the first letter of the second name of each constellation is incorporated. Finally, mention must be made of Crater (CRT) which has been abbreviated in such a way as to avoid confusion with the aforementioned CRA (Corona Australis).

The table shown on the following pages contains the name of each of the 88 constellations together with the translation and abbreviation of the constellation name. The constellations depicted on the monthly star charts are identified with their abbreviations rather than the full constellation names.

The Constellations

Andromeda	Andromeda	AND		Delphinus	The Dolphin	DEL
Antlia	The Air Pump	ANT		Dorado	The Goldfish	DOR
Apus	The Bird of Paradise	APS		Draco	The Dragon	DRA
				Equuleus	The Foal	EQU
Aquarius	The Water Carrier	AQR		Eridanus	The River	ERI
Aquila	The Eagle	AQL		Fornax	The Furnace	FOR
Ara	The Altar	ARA		Gemini	The Twins	GEM
Aries	The Ram	ARI		Grus	The Crane	GRU
Auriga	The Charioteer	AUR		Hercules	Hercules	HER
Boötes	The Herdsman	BOO		Horologium	The Pendulum Clock	HOR
Caelum	The Graving Tool	CAE				
Camelopardalis	The Giraffe	CAM		Hydra	The Water Snake	HYA
Cancer	The Crab	CNC		Hydrus	The Lesser Water Snake	HYI
Canes Venatici	The Hunting Dogs	CVN				
Canis Major	The Great Dog	CMA		Indus	The Indian	IND
Canis Minor	The Little Dog	CMI		Lacerta	The Lizard	LAC
Capricornus	The Goat	CAP		Leo	The Lion	LEO
Carina	The Keel	CAR		Leo Minor	The Lesser Lion	LMI
Cassiopeia	Cassiopeia	CAS		Lepus	The Hare	LEP
Centaurus	The Centaur	CEN		Libra	The Scales	LIB
Cepheus	Cepheus	CEP		Lupus	The Wolf	LUP
Cetus	The Whale	CET		Lynx	The Lynx	LYN
Chamaeleon	The Chameleon	CHA		Lyra	The Lyre	LYR
Circinus	The Pair of Compasses	CIR		Mensa	The Table Mountain	MEN
				Microscopium	The Microscope	MIC
Columba	The Dove	COL		Monoceros	The Unicorn	MON
Coma Berenices	Berenice's Hair	COM		Musca	The Fly	MUS
Corona Australis	The Southern Crown	CRA		Norma	The Level	NOR
Corona Borealis	The Northern Crown	CRB		Octans	The Octant	OCT
				Ophiuchus	The Serpent Bearer	OPH
Corvus	The Crow	CRV		Orion	Orion	ORI
Crater	The Cup	CRT		Pavo	The Peacock	PAV
Crux	The Cross	CRU		Pegasus	Pegasus	PEG
Cygnus	The Swan	CYG		Perseus	Perseus	PER

Phoenix	The Phoenix	PHE		Sextans	The Sextant	SEX
Pictor	The Painter's Easel	PIC		Taurus	The Bull	TAU
Pisces	The Fish	PSC		Telescopium	The Telescope	TEL
Piscis Austrinus	The Southern Fish	PSA		Triangulum	The Triangle	TRI
Puppis	The Stern	PUP		Triangulum Australe	The Southern Triangle	TRA
Pyxis	The Mariner's Compass	PYX		Tucana	The Toucan	TUC
Reticulum	The Net	RET		Ursa Major	The Great Bear	UMA
Sagitta	The Arrow	SGE		Ursa Minor	The Little Bear	UMI
Sagittarius	The Archer	SGR		Vela	The Sail	VEL
Scorpius	The Scorpion	SCO		Virgo	The Virgin	VIR
Sculptor	The Sculptor	SCL		Volans	The Flying Fish	VOL
Scutum	The Shield	SCT		Vulpecula	The Fox	VUL
Serpens Caput and Cauda	The Serpent	SER				

The Monthly Star Charts

Northern Hemisphere Star Charts

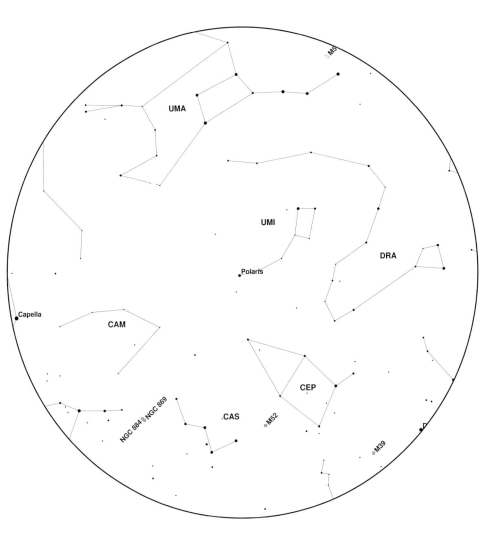

This chart shows stars lying at declinations between +45 and +90 degrees. These constellations are circumpolar for observers in Europe and North America.

1N

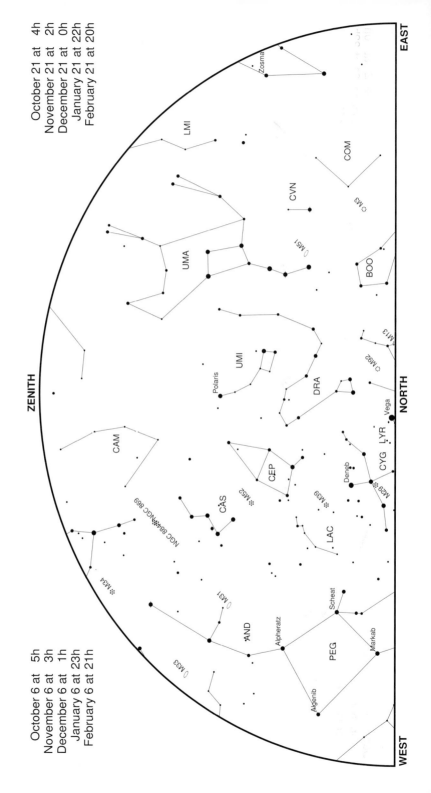

October 6 at 5h
November 6 at 3h
December 6 at 1h
January 6 at 23h
February 6 at 21h

EAST

ZENITH

NORTH

WEST

Zosma

LMI

COM

CVN

M3

BOO

UMA

M51

M13

M92

UMI

Polaris

DRA

LYR

Vega

CAM

CEP

CYG

M52

Deneb

CAS

M29

M39

NGC 884/NGC 869

LAC

M34

M31

AND

Alpheratz

Scheat

PEG

M33

Markab

Algenib

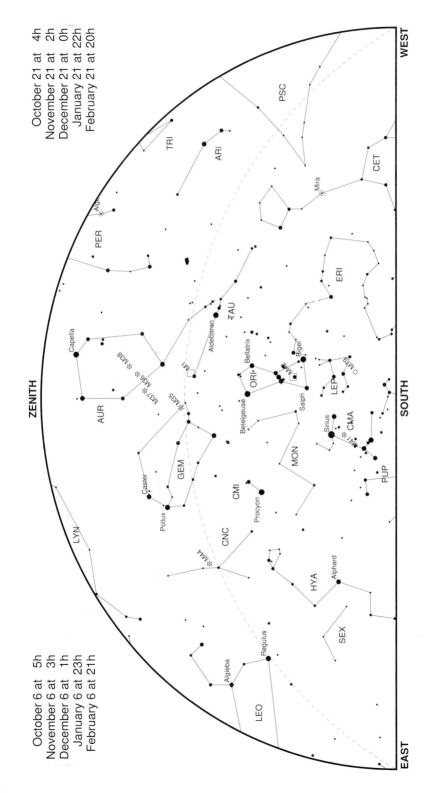

1S

WEST

PSC

CET

Mira

TRI

ARI

PER

Algol

ERI

Capella

TAU

Aldebaran

M38

M36

M37

M35

AUR

M1

Bellatrix

Rigel

ORI

M42

M79

Betelgeuse

Saiph

LEP

Castor

GEM

MON

Sirius

CMA

Pollux

M41

Procyon

CMI

PUP

ZENITH

LYN

CNC

M44

SOUTH

HYA

Alphard

SEX

Algieba

Regulus

LEO

EAST

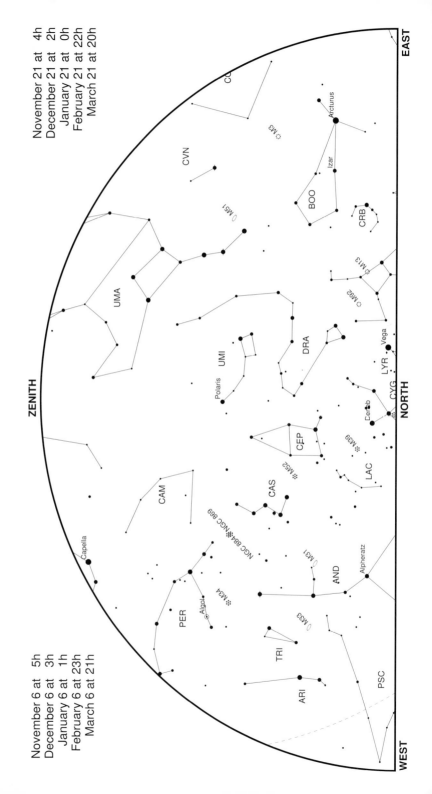

2N

November 6 at 5h
December 6 at 3h
January 6 at 1h
February 6 at 23h
March 6 at 21h

ZENITH

EAST

NORTH

WEST

Arcturus

Izar

BOO

CRB

M3

CO

CVN

M51

UMA

M13

M92

Vega

LYR

CYG

DRA

Polaris

UMI

Deneb

M39

LAC

CEP

M52

CAS

CAM

NGC 884·❋·NGC 869

Capella

M34

Algol

PER

TRI

ARI

M33

M31

AND

Alpheratz

PSC

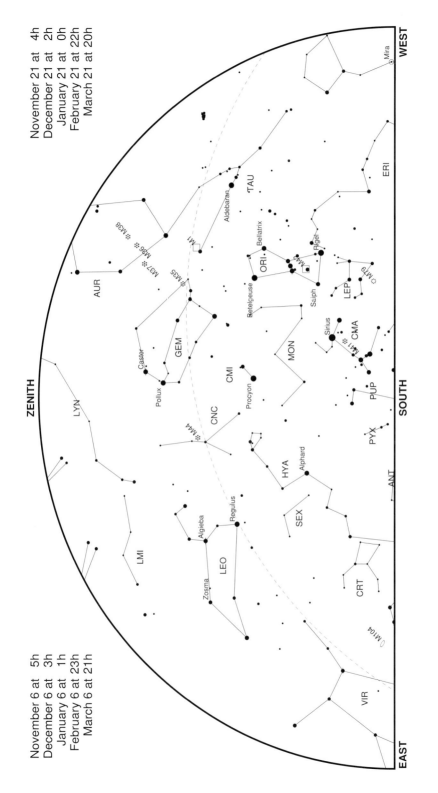

2S

WEST

SOUTH

EAST

ZENITH

November 21 at 4h
December 21 at 2h
January 21 at 0h
February 21 at 22h
March 21 at 20h

November 6 at 5h
December 6 at 3h
January 6 at 1h
February 6 at 23h
March 6 at 21h

Mira

ERI

TAU
Aldebaran
M1
ORI
Bellatrix
Rigel
M42
Betelgeuse
Saiph
LEP
M79
Sirius
CMA
M41

AUR
M38
M36
M37
M35
GEM
Castor
Pollux
CMI
Procyon
MON
PUP
PYX
ANT

CNC
M44

LYN
LMI

LEO
Algieba
Regulus
Zosma

HYA
Alphard
SEX
CRT
M104
VIR

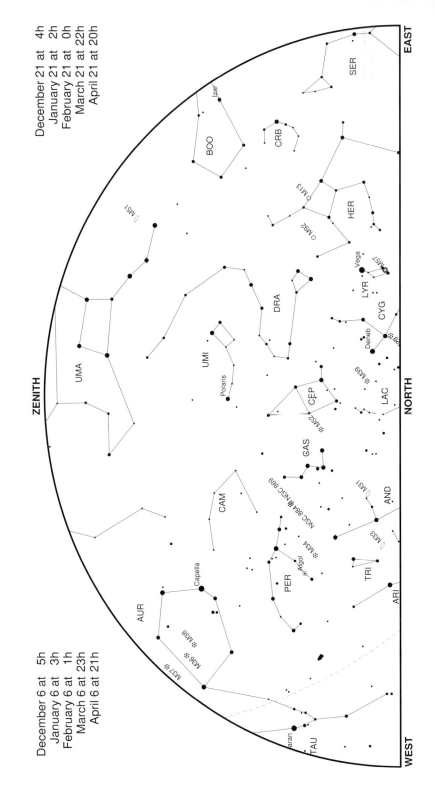

3N

December 21 at 4h
January 21 at 2h
February 21 at 0h
March 21 at 22h
April 21 at 20h

December 6 at 5h
January 6 at 3h
February 6 at 1h
March 6 at 23h
April 6 at 21h

EAST

SER

BOO

Izar

CRB

HER

M13

M92

LYR

Vega

M57

CYG

Deneb

M51

DRA

UMI

Polaris

UMA

ZENITH

M39

LAC

M52

CEP

GAS

CAM

NGC 884

NGC 869

M34

Algol

PER

M31

AND

M33

TRI

ARI

Capella

AUR

M38

M36

M37

TAU

aran

WEST

NORTH

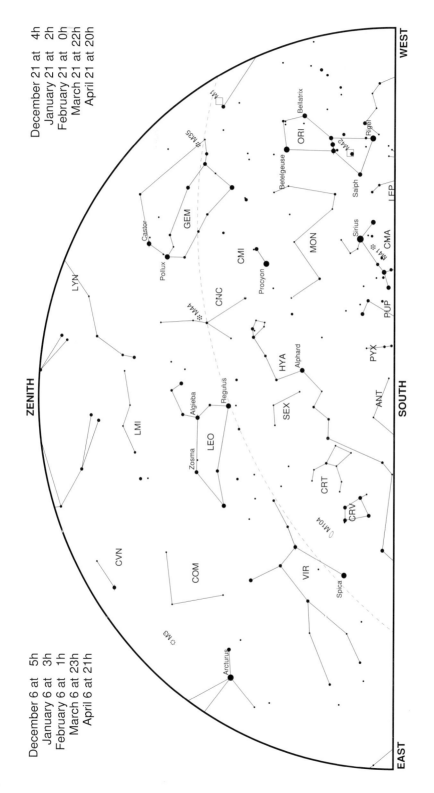

3S

WEST

EAST

ZENITH

SOUTH

Bellatrix
ORI
Rigel
Betelgeuse
M42
Salph
LEP
Sirius
CMA
M41
PUP
PYX
ANT
Procyon
CMI
MON
CNC
M44
GEM
Castor
Pollux
M35
LYN
HYA
Alphard
SEX
LMI
Algieba
Regulus
LEO
Zosma
CRT
CRV
M104
VIR
Spica
COM
CVN
M3
Arcturus

4N

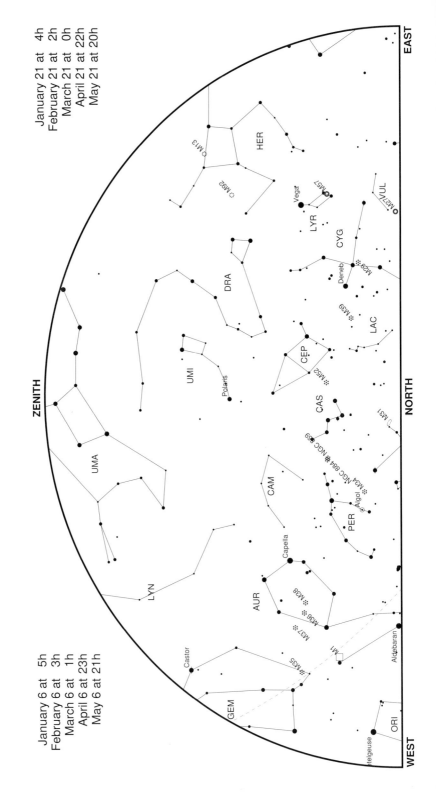

January 21 at 4h
February 21 at 2h
March 21 at 0h
April 21 at 22h
May 21 at 20h

January 6 at 5h
February 6 at 3h
March 6 at 1h
April 6 at 23h
May 6 at 21h

EAST

ZENITH

NORTH

WEST

HER
M13
M92
Vega
M57
M27/UL
LYR
CYG
Deneb
M29
M39
LAC
DRA
CEP
M52
UMI
Polaris
CAS
M31
UMA
M34 NGC 884 NGC 869
Algol
CAM
PER
Capella
AUR
M38
M36
M37
LYN
M1
Aldebaran
Castor
GEM
M35
ORI
Betelgeuse

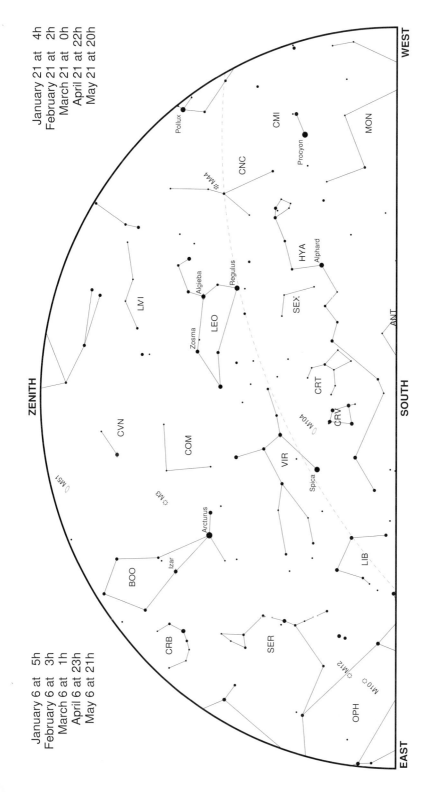

4S

5N

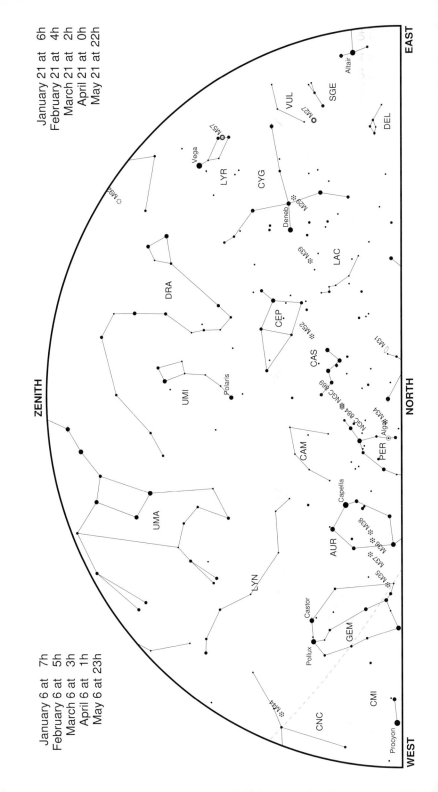

January 21 at 6h
February 21 at 4h
March 21 at 2h
April 21 at 0h
May 21 at 22h

January 6 at 7h
February 6 at 5h
March 6 at 3h
April 6 at 1h
May 6 at 23h

EAST

NORTH

WEST

ZENITH

Altair

SGE

VUL

M27

DEL

Vega

M57

LYR

CYG

Deneb

M29

M39

LAC

CEP

M52

CAS

M31

NGC 884 NGC 869

M34

Algol

PER

Capella

AUR

M38

M36

M37

M35

GEM

Castor

Pollux

CMI

Procyon

CNC

M44

LYN

UMA

CAM

UMI

Polaris

DRA

M92

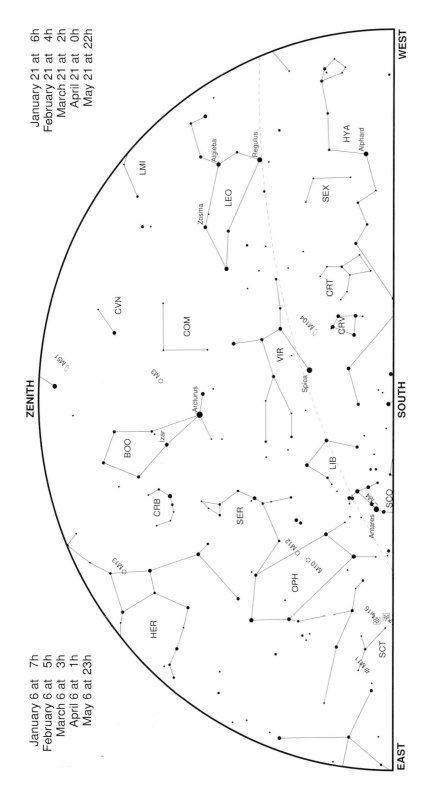

5S

WEST

January 21 at 6h
February 21 at 4h
March 21 at 2h
April 21 at 0h
May 21 at 22h

January 6 at 7h
February 6 at 5h
March 6 at 3h
April 6 at 1h
May 6 at 23h

ZENITH

EAST

SOUTH

6N

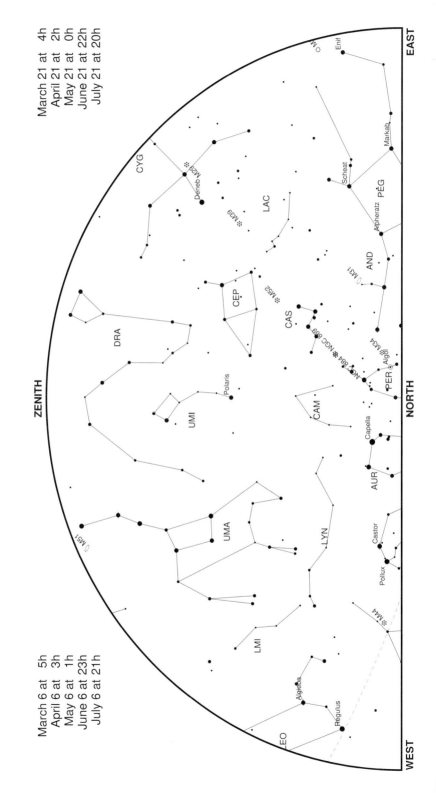

March 6 at 5h
April 6 at 3h
May 6 at 1h
June 6 at 23h
July 6 at 21h

ZENITH

EAST

NORTH

WEST

CYG
Deneb
M29
M39
LAC
CEP
M52
DRA
CAS
UMI
Polaris
NGC 869
NGC 884
M34
Algol
PER
CAM
Capella
AUR
M81
UMA
LYN
LMI
Castor
Pollux
M44
LEO
Algieba
Regulus
Enif
Markab
Scheat
PEG
Alpheratz
AND
M31

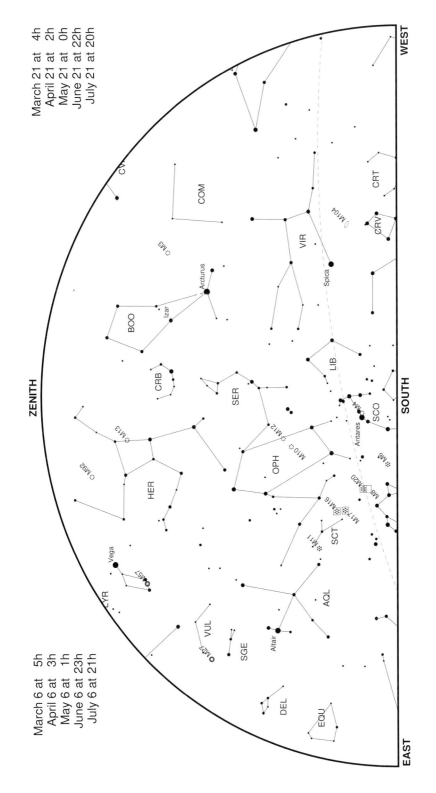

6S

WEST

March 21 at 4h
April 21 at 2h
May 21 at 0h
June 21 at 22h
July 21 at 20h

ZENITH

CVn

COM

CRT

CRV

M104

VIR

Spica

BOO

Arcturus

Izar

M3

CRB

SER

LIB

M92

M13

HER

OPH

M12

M10

M4

SCO

Antares

M6

M9

M20

M8

M16

M17

M11

SCT

SOUTH

LYR

Vega

M57

VUL

M27

SGE

Altair

AQL

DEL

EQU

March 6 at 5h
April 6 at 3h
May 6 at 1h
June 6 at 23h
July 6 at 21h

EAST

7N

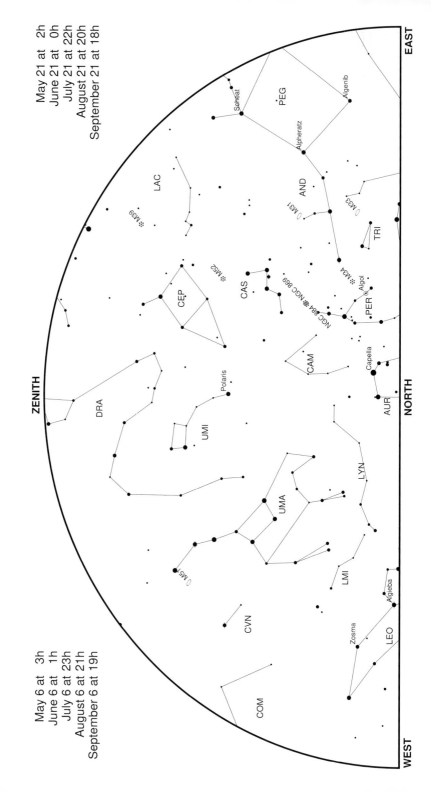

May 21 at 2h
June 21 at 0h
July 21 at 22h
August 21 at 20h
September 21 at 18h

May 6 at 3h
June 6 at 1h
July 6 at 23h
August 6 at 21h
September 6 at 19h

EAST

ZENITH

NORTH

WEST

Scheat
PEG
Algenib
Alpheratz
AND
M31
M33
LAC
M39
TRI
M52
Algol
CAS
PER
NGC 884 # NGC 869
CEP
CAM
Capella
AUR
DRA
Polaris
UMI
LYN
UMA
LMI
M51
CVN
Algieba
LEO
COM
Zosma

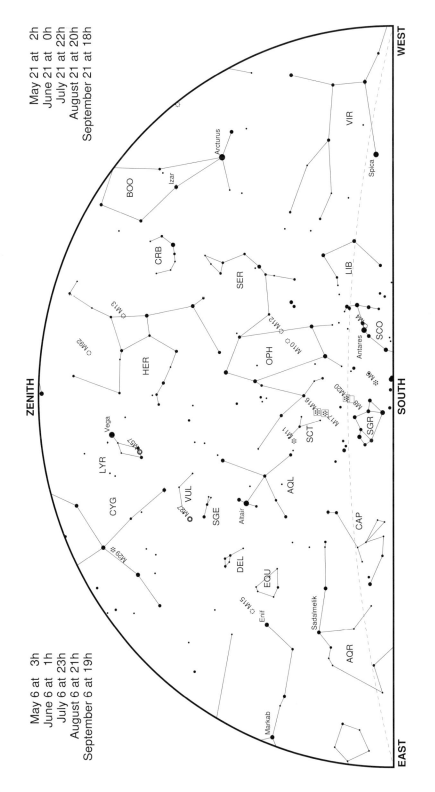

7S

WEST

May 21 at 2h
June 21 at 0h
July 21 at 22h
August 21 at 20h
September 21 at 18h

ZENITH

VIR

Spica

BOO

Izar

Arcturus

CRB

SER

LIB

M13

HER

M92

OPH

M10

M12

SCO

Antares

M4

LYR

Vega

M57

CYG

M29

VUL

M27

SGE

SCT

M11

M16

M17

M18

M20

M8

SGR

M6

SOUTH

AQL

Altair

DEL

CAP

EQU

M15

Enif

Sadalmelik

AQR

Markab

May 6 at 3h
June 6 at 1h
July 6 at 23h
August 6 at 21h
September 6 at 19h

EAST

8N

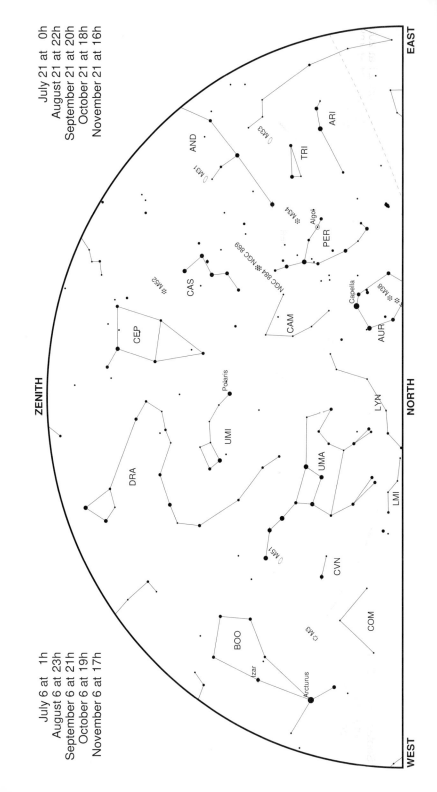

ZENITH

EAST

July 21 at 0h
August 21 at 22h
September 21 at 20h
October 21 at 18h
November 21 at 16h

July 6 at 1h
August 6 at 23h
September 6 at 21h
October 6 at 19h
November 6 at 17h

WEST

NORTH

AND

M33

ARI

TRI

M34

Algol

PER

NGC 884

NGC 869

M52

CAS

CAM

Capella

M38

AUR

CEP

Polaris

LYN

UMI

DRA

UMA

LMI

M51

CVN

COM

M3

BOO

Izar

Arcturus

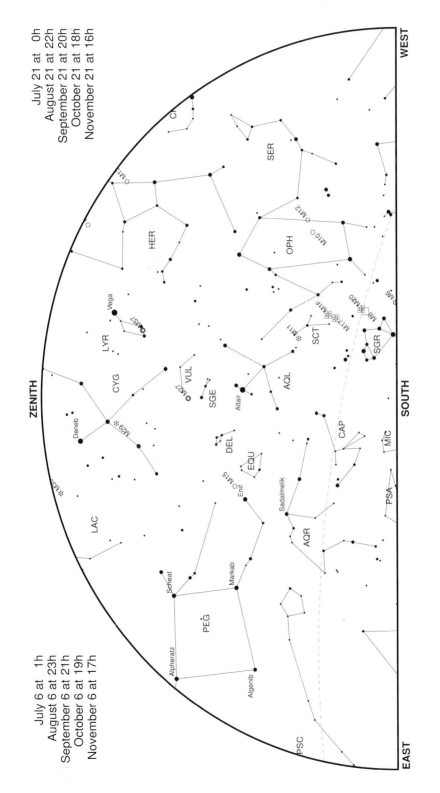

8S

WEST

ZENITH

SOUTH

EAST

July 21 at 0h
August 21 at 22h
September 21 at 20h
October 21 at 18h
November 21 at 16h

July 6 at 1h
August 6 at 23h
September 6 at 21h
October 6 at 19h
November 6 at 17h

CY
SER
M13
HER
OPH
M12
M10
Vega
M57
LYR
M11
SCT
M16 M17 M18
M20 M8
CYG
M29
Deneb
VUL
M27
SGE
Altair
AQL
SGR
M22
M8
M39
LAC
DEL
EQU
M15
Enif
CAP
MIC
Sadalmelik
AQR
PSA
Scheat
Markab
PEG
Alpheratz
Algenib
PSC

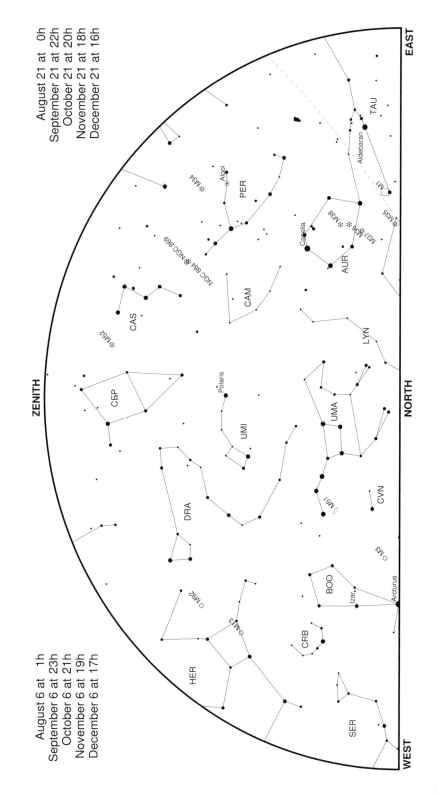

9N

August 21 at 0h
September 21 at 22h
October 21 at 20h
November 21 at 18h
December 21 at 16h

August 6 at 1h
September 6 at 23h
October 6 at 21h
November 6 at 19h
December 6 at 17h

EAST
ZENITH
NORTH
WEST

TAU
Aldebaran
M1
M35
M37 M36 M38
AUR
Capella
PER
Algol
M34
CAM
NGC 884 NGC 869
CAS
M52
CEP
Polaris
UMI
LYN
UMA
M51
CVN
M3
DRA
M92
BOO
Izar
Arcturus
CRB
M13
HER
SER

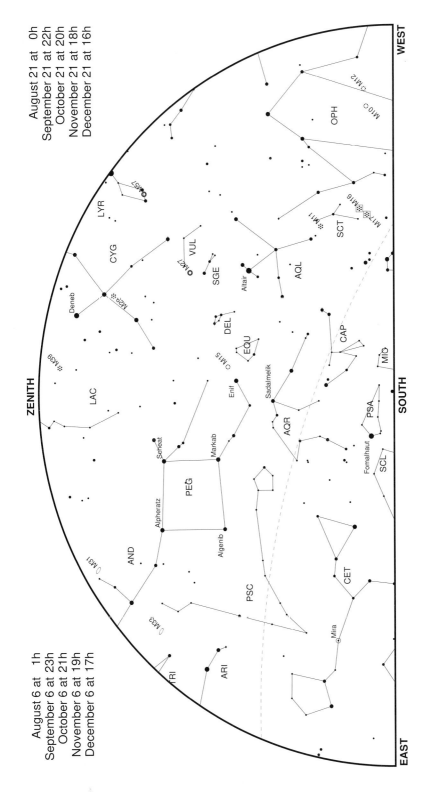

9S

WEST

August 21 at 0h
September 21 at 22h
October 21 at 20h
November 21 at 18h
December 21 at 16h

ZENITH

August 6 at 1h
September 6 at 23h
October 6 at 21h
November 6 at 19h
December 6 at 17h

EAST

SOUTH

OPH
M12
M10
M16 M17
SCT
M11
AQL
Altair
VUL
M27
SGE
LYR
M57
CYG
Deneb
M29
M39
LAC
DEL
EQU
M15
Enif
CAP
Mi
PSA
Fomalhaut
SCL
AQR
Sadalmelik
PEG
Scheat
Markab
Alpheratz
Algenib
AND
M31
M33
TRI
ARI
PSC
CET
Mira

10N

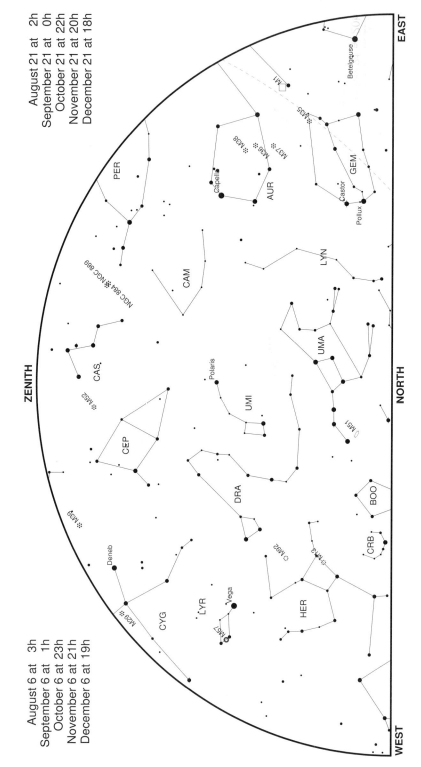

EAST

PER

CAM

M38
M36
M37
M35

Capella

AUR

GEM

Castor

Pollux

Betelgeuse

LYN

NGC 884
NGC 869

CAS.

M52

Polaris

UMI

UMA

M51

NORTH

ZENITH

CEP

DRA

BOO

M29

M92

M13

CRB

Deneb

CYG

LYR

Vega

M57

HER

WEST

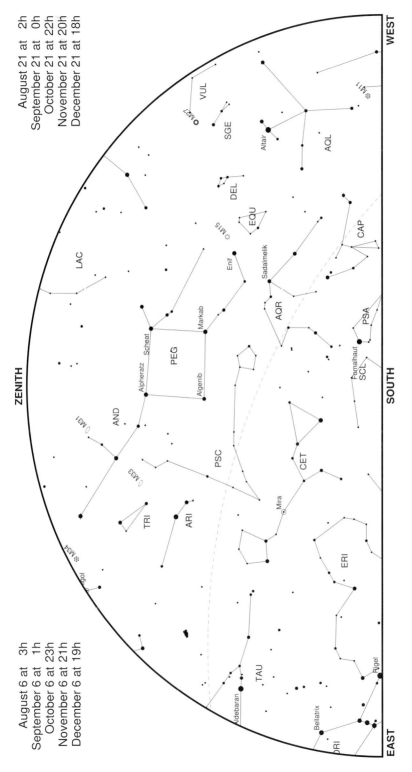

10S

WEST

August 21 at 2h
September 21 at 0h
October 21 at 22h
November 21 at 20h
December 21 at 18h

August 6 at 3h
September 6 at 1h
October 6 at 23h
November 6 at 21h
December 6 at 19h

ZENITH

EAST

SOUTH

VUL
M27
SGE
Altair
AQL
M71
DEL
EQU
M15
Enif
Sadalmelik
AQR
CAP
LAC
Markab
Scheat
Alpheratz
PEG
Algenib
PSA
Fomalhaut
SCL
AND
M31
M33
PSC
CET
Mira
TRI
ARI
M34
Algol
ERI
TAU
Aldebaran
Rigel
Bellatrix
ORI

11N

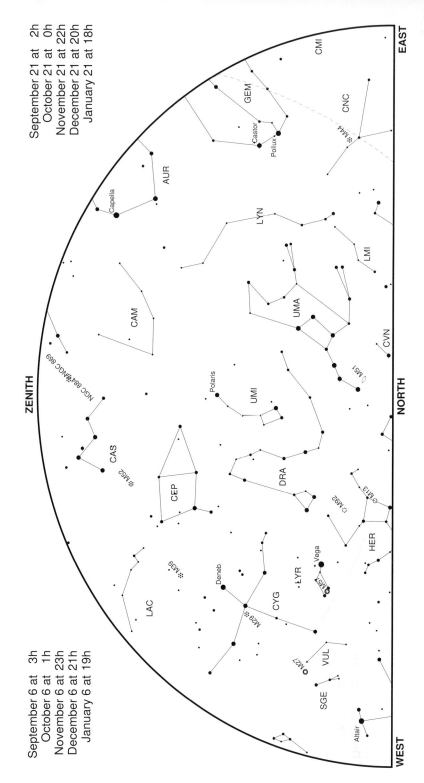

September 6 at 3h
October 6 at 1h
November 6 at 23h
December 6 at 21h
January 6 at 19h

EAST

ZENITH

NORTH

WEST

CMI
GEM
Castor
Pollux
CNC
M44
AUR
Capella
LYN
LMI
CAM
UMA
CVN
M51
Polaris
UMI
NGC 884 NGC 869
CAS
M52
CEP
DRA
M92
M13
HER
M39
Deneb
M57
Vega
LYR
LAC
M29
CYG
M27
VUL
SGE
Altair

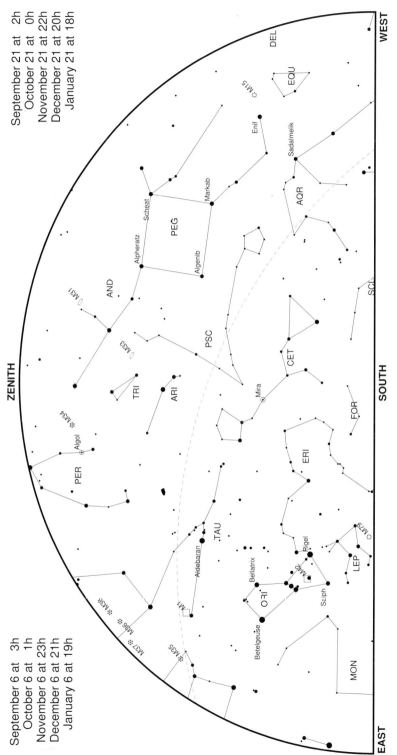

11S

September 21 at 2h
October 21 at 0h
November 21 at 22h
December 21 at 20h
January 21 at 18h

September 6 at 3h
October 6 at 1h
November 6 at 23h
December 6 at 21h
January 6 at 19h

ZENITH

WEST

EAST

SOUTH

DEL
EQU
M15
Enif
Sadalmelik
Markab
Scheat
PEG
Alpheratz
Algenib
AQR
AND
M31
PSC
SCL
M33
CET
TRI
ARI
Mira
FOR
Algol
M34
PER
ERI
M79
LEP
Rigel
TAU
Aldebaran
Bellatrix
M42
Saiph
M1
ORI
Betelgeuse
MON
M36
M37
M35

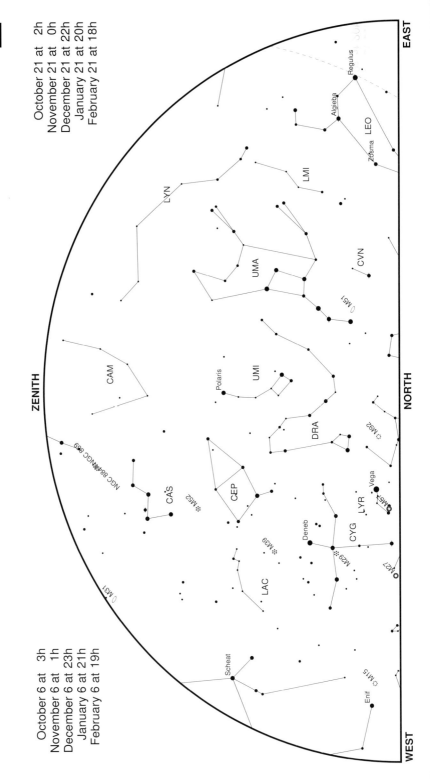

12N

October 21 at 2h
November 21 at 0h
December 21 at 22h
January 21 at 20h
February 21 at 18h

October 6 at 3h
November 6 at 1h
December 6 at 23h
January 6 at 21h
February 6 at 19h

EAST

WEST

NORTH

ZENITH

LEO · Regulus · Algieba · Zosma
LMI
CVN · M51
UMA
LYN
CAM
UMI · Polaris
DRA
M92
CAS · M52 · NGC 884 · NGC 869
CEP
LYR · Vega · M57
CYG · Deneb · M29
M39
LAC
M31
M27
M15
Scheat · Enif

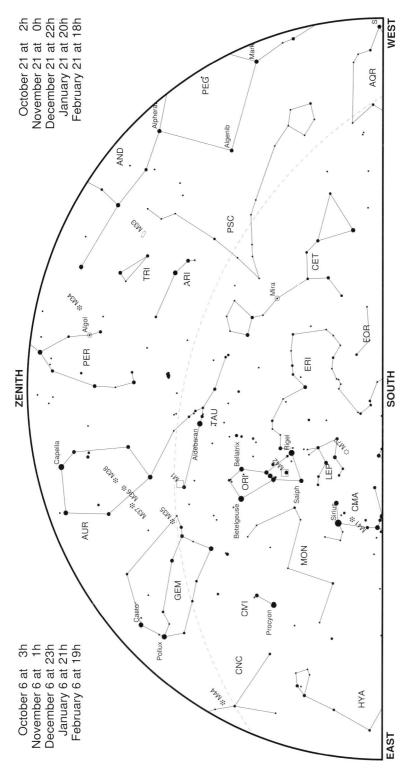

12S

October 21 at 2h
November 21 at 0h
December 21 at 22h
January 21 at 20h
February 21 at 18h

October 6 at 3h
November 6 at 1h
December 6 at 23h
January 6 at 21h
February 6 at 19h

ZENITH

WEST

EAST

SOUTH

AQR

PEG
Markab
Alpheratz
AND
Algenib
PSC
CET
Mira
ERI
EOR
TRI
M33
ARI
PER
Algol
M34

Capella
AUR
M38
M36
M37
M35
M1
TAU
Aldebaran
Bellatrix
ORI
Betelgeuse
M42
Rigel
Saiph
LEP
M79
CMA
Sirius
M41

GEM
Castor
Pollux
CNC
M44
CMI
Procyon
MON
HYA

Southern Hemisphere Star Charts

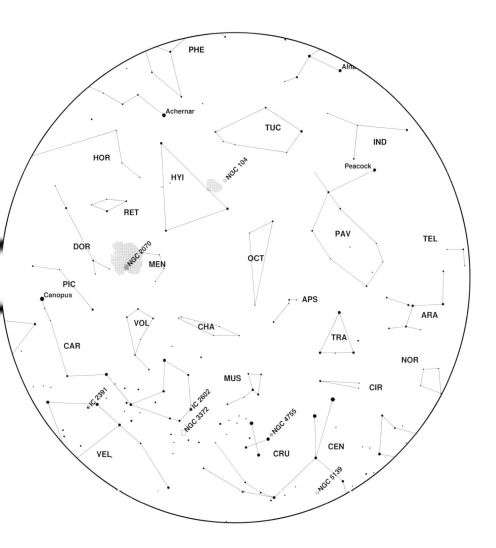

This chart shows stars lying at declinations between − 45 and − 90 degrees. These constellations are circumpolar for observers in Australia and New Zealand.

1N

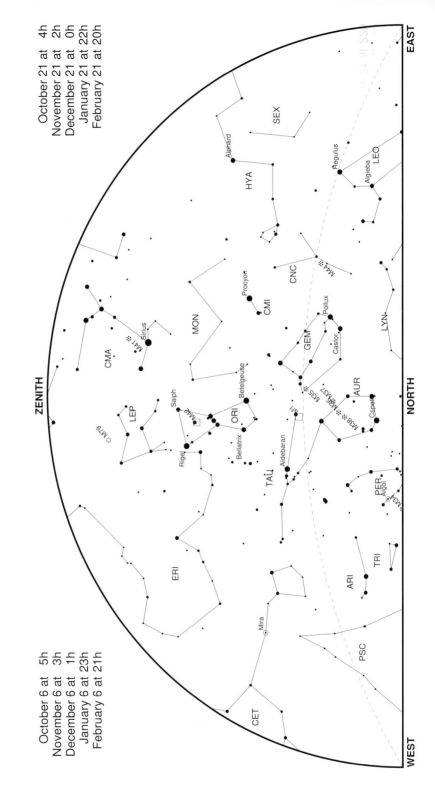

October 21 at 4h
November 21 at 2h
December 21 at 0h
January 21 at 22h
February 21 at 20h

October 6 at 5h
November 6 at 3h
December 6 at 1h
January 6 at 23h
February 6 at 21h

EAST

ZENITH

NORTH

WEST

SEX

HYA

Alphard

Regulus

LEO

Algieba

CNC

M44

Procyon

CMI

MON

GEM

Pollux

Castor

LYN

Sirius

M41

CMA

AUR

M35

M36 M37

Capella

Betelgeuse

M38

LEP

Saiph

M42

ORI

M79

Rigel

Bellatrix

M1

LYN

Aldebaran

PER

Algol

TAU

Mirb

ERI

TRI

ARI

Mira

PSC

CET

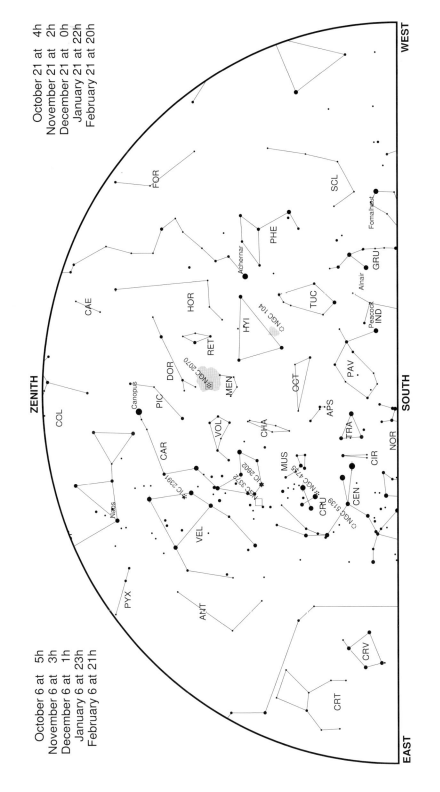

1S

WEST

October 21 at 4h
November 21 at 2h
December 21 at 0h
January 21 at 22h
February 21 at 20h

October 6 at 5h
November 6 at 3h
December 6 at 1h
January 6 at 23h
February 6 at 21h

ZENITH

EAST

SOUTH

FOR
SCL
Achernar
PHE
CAE
TUC
NGC 104
HOR
GRU
Alnair
HYI
Peacock
IND
RET
MEN
OCT
PAV
DOR
NGC 2070
PIC
APS
COL
Canopus
VOL
CHA
TRA
CAR
CEN
NOR
IC 2391
NGC 3372
IC 2602
MUS
CIR
NGC 5139
CRU
NGC 4755
VEL
Naos
PYX
ANT
CRT
CRV
Fomalhaut

2N

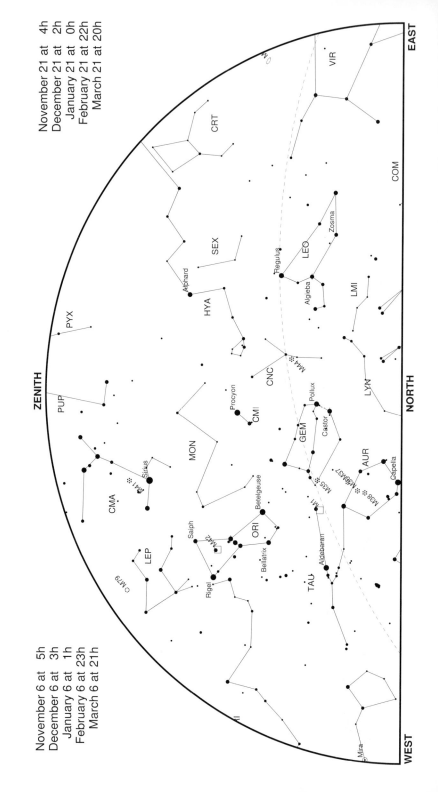

EAST

ZENITH

NORTH

WEST

November 6 at 5h
December 6 at 3h
January 6 at 1h
February 6 at 23h
March 6 at 21h

VIR
CRT
SEX
COM
LEO
Zosma
Algieba
Regulus
LMI
HYA
Alphard
PYX
CNC
M44
PUP
Procyon
CMI
LYN
MON
GEM
Pollux
Castor
Sirius
M41
CMA
M35
AUR
M38 M36 M37
Capella
Saiph
M42
Betelgeuse
ORI
LEP
Bellatrix
M79
Rigel
MI
Aldebaran
TAU
Mira

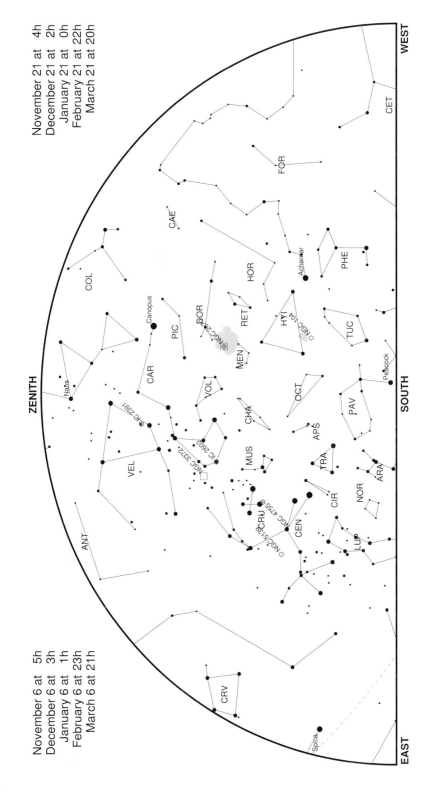

2S

WEST

November 21 at 4h
December 21 at 2h
January 21 at 0h
February 21 at 22h
March 21 at 20h

November 6 at 5h
December 6 at 3h
January 6 at 1h
February 6 at 23h
March 6 at 21h

ZENITH

EAST

SOUTH

CET
FOR
CAE
PHE
Achernar
COL
HOR
RET
TUC
NGC 104
HYI
Canopus
PIC
DOR
NGC 2070
MEN
Nabs
CAR
VOL
OCT
PAV
Peacock
IC 2391
VEL
CHA
APS
NGC 3372
IC 2602
MUS
TRA
ARA
ANT
NGC 5138
NGC 4755
CRU
CEN
CIR
NOR
LUP
CRV
Spica

3N

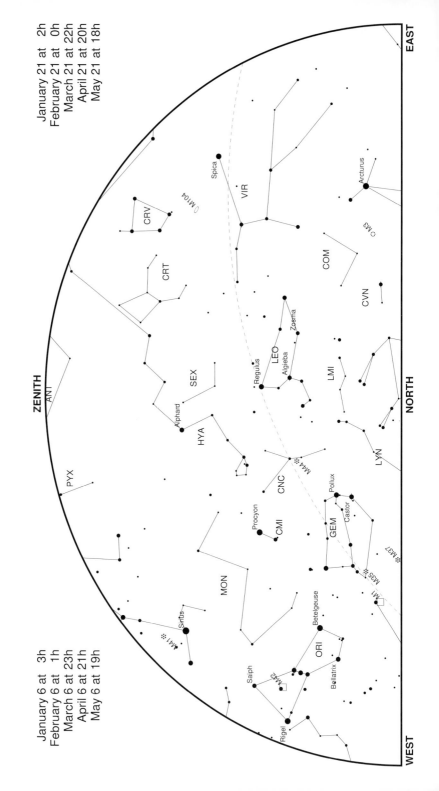

January 21 at 2h
February 21 at 0h
March 21 at 22h
April 21 at 20h
May 21 at 18h

ZENITH

EAST

January 6 at 3h
February 6 at 1h
March 6 at 23h
April 6 at 21h
May 6 at 19h

WEST

NORTH

Spica

VIR

Arcturus

M3

CRV

M104

CRT

COM

CVN

SEX

Regulus

LEO

Zosma

Algieba

LMI

ANT

Alphard

HYA

PYX

LYN

Procyon

CMI

CNC

M44

Pollux

GEM

Castor

M37

MON

M35

M41

Sirius

M1

Betelgeuse

ORI

Saiph

M42

Bellatrix

Rigel

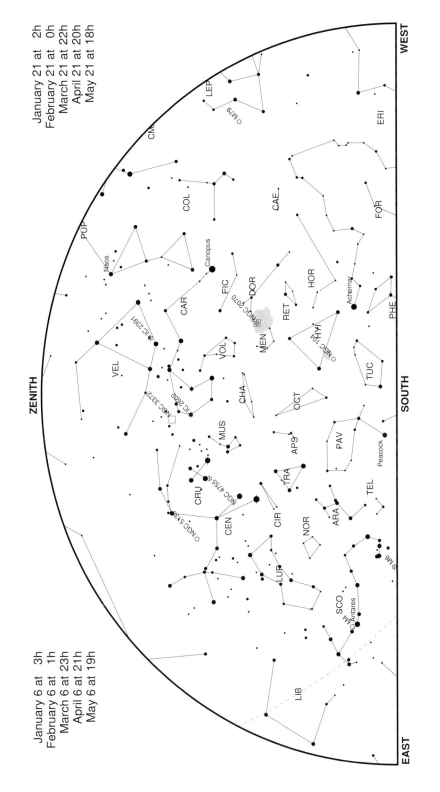

3S

WEST

EAST

ZENITH

SOUTH

January 21 at 2h
February 21 at 0h
March 21 at 22h
April 21 at 20h
May 21 at 18h

January 6 at 3h
February 6 at 1h
March 6 at 23h
April 6 at 21h
May 6 at 19h

LEP
M79
ERI
CMA
COL
CAE
FOR
PUP
Naos
Canopus
CAR
FIC
DOR
HOR
Achernar
PHE
NGC 2070
RET
MEN
HYI
NGC 104
TUC
IC 2391
VEL
VOL
NGC 3372
IC 2602
CHA
OCT
NGC 2391
MUS
APS
PAV
Peacock
CRU
TRA
NGC 4755
CEN
CIR
TEL
NGC 5139
NOR
ARA
LUP
M6
SCO
M4
Antares
LIB

4N

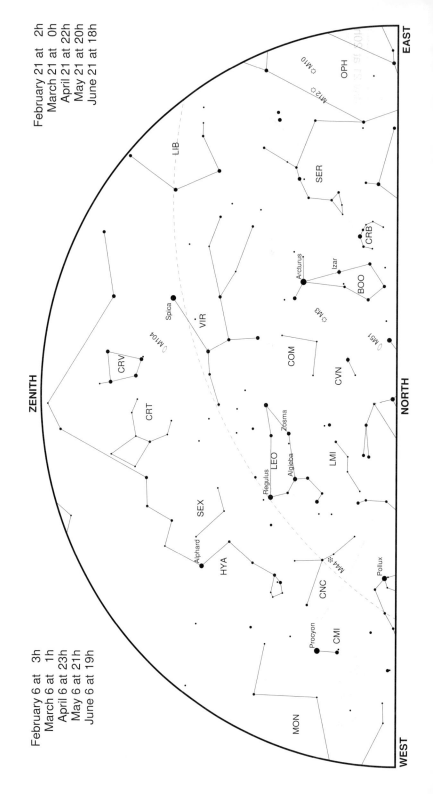

February 21 at 2h
March 21 at 0h
April 21 at 22h
May 21 at 20h
June 21 at 18h

February 6 at 3h
March 6 at 1h
April 6 at 23h
May 6 at 21h
June 6 at 19h

EAST

WEST

ZENITH

NORTH

OPH
M10
M12
LIB
SER
CRB
Arcturus
Izar
BOO
M3
Spica
M104
VIR
CRV
COM
CRT
CVN
M51
Zosma
LEO
Algieba
LMI
Regulus
SEX
Alphard
HYA
M44
CNC
Pollux
Procyon
CMI
MON

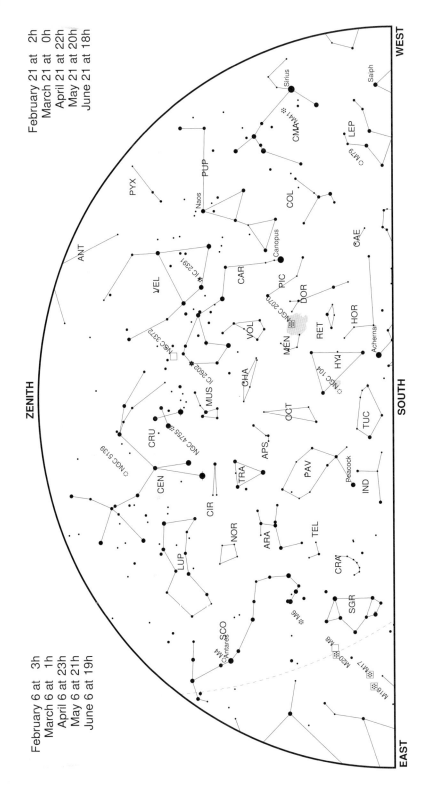

4S

WEST

February 21 at 2h
March 21 at 0h
April 21 at 22h
May 21 at 20h
June 21 at 18h

February 6 at 3h
March 6 at 1h
April 6 at 23h
May 6 at 21h
June 6 at 19h

ZENITH

EAST

SOUTH

5N

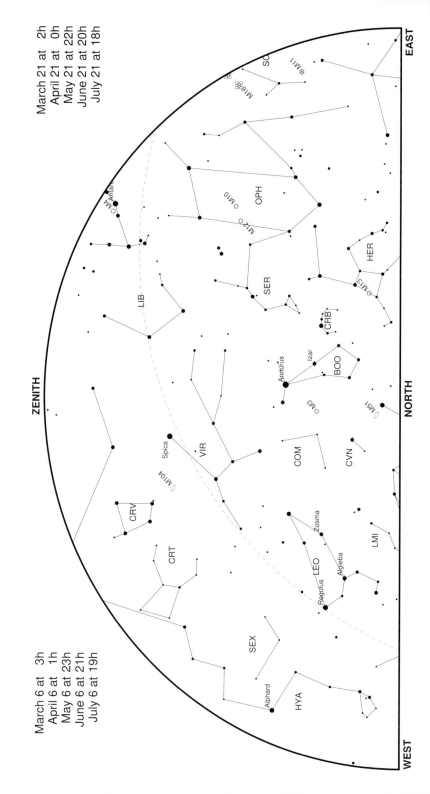

March 21 at 2h
April 21 at 0h
May 21 at 22h
June 21 at 20h
July 21 at 18h

March 6 at 3h
April 6 at 1h
May 6 at 23h
June 6 at 21h
July 6 at 19h

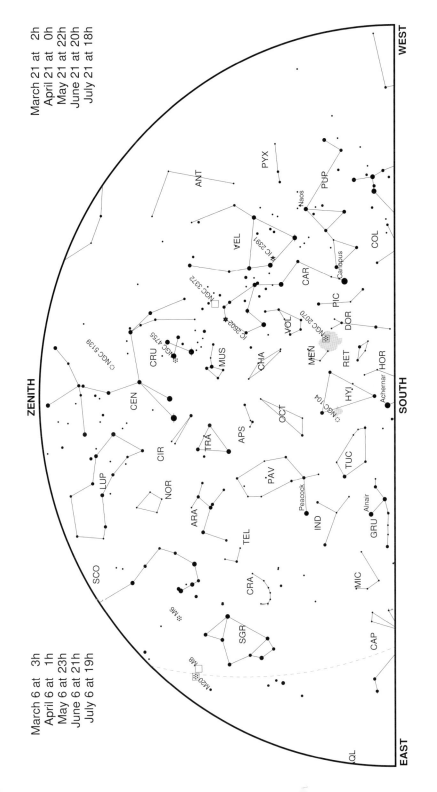

5S

WEST

EAST

ZENITH

SOUTH

March 21 at 2h
April 21 at 0h
May 21 at 22h
June 21 at 20h
July 21 at 18h

March 6 at 3h
April 6 at 1h
May 6 at 23h
June 6 at 21h
July 6 at 19h

ANT
PYX
PUP
Naos
VEL
IC 2391
CAR
Canopus
COL
NGC 3372
NGC 5139
IC 2602
PIC
VOL
NGC 2070
DOR
CRU
NGC 4755
MUS
CHA
MEN
RET
HOR
CEN
Achernar
HYI
NGC 104
CIR
APS
OCT
TUC
TRA
LUP
PAV
NOR
Peacock
IND
ARA
GRU
Alnair
TEL
SCO
CRA
MIC
M6
SGR
CAP
M20
M8

6N

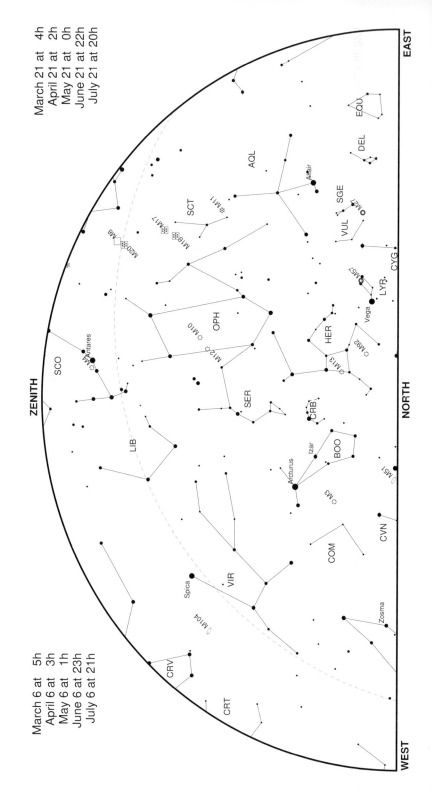

March 6 at 5h
April 6 at 3h
May 6 at 1h
June 6 at 23h
July 6 at 21h

EAST

ZENITH

NORTH

WEST

EQU

DEL

SGE

VUL

CYG

LYR.

Vega

M57

M92

HER

M13

M51

CVN

CRB

BOO

Izar

Arcturus

M3

Zosma

COM

OPH

M10

M12

SER

SCO

Antares

LIB

VIR

Spica

M104

CRV

CRT

AQL

SCT

M11

M17

M16

M20

M8

Altair

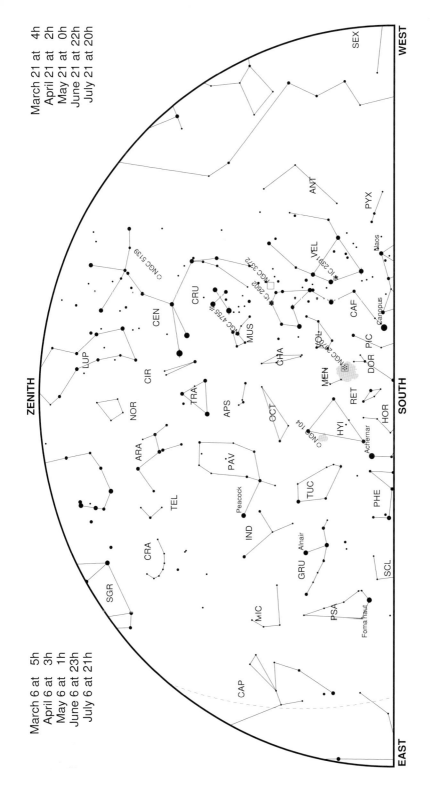

6S

WEST

SEX

March 21 at 4h
April 21 at 2h
May 21 at 0h
June 21 at 22h
July 21 at 20h

ZENITH

ANT

PYX

NGC 5139

CEN

VEL

IC 2391

CRU

NGC 3372

C 262

CAF

NGC 4755

MUS

LUP

PIC

CIR

CHA

VOL

NGC 2070

DOR

MEN

NOR

TRA

RET

APS

OCT

ARA

HYI

NGC 104

TEL

PAV

HOR

Achernar

SOUTH

CRA

Peacock

TUC

PHE

SGR

IND

GRU   Alnair

SCL

MIC

PSA

Fomalhaut

CAP

EAST

March 6 at 5h
April 6 at 3h
May 6 at 1h
June 6 at 23h
July 6 at 21h

Naos

Canopus

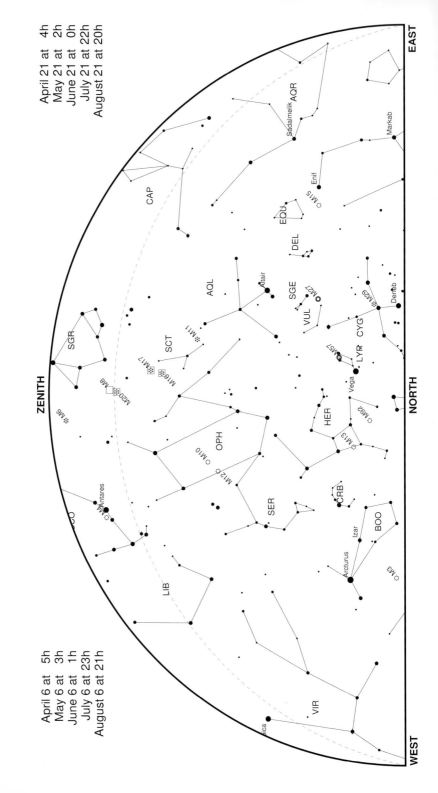

April 21 at 4h
May 21 at 2h
June 21 at 0h
July 21 at 22h
August 21 at 20h

April 6 at 5h
May 6 at 3h
June 6 at 1h
July 6 at 23h
August 6 at 21h

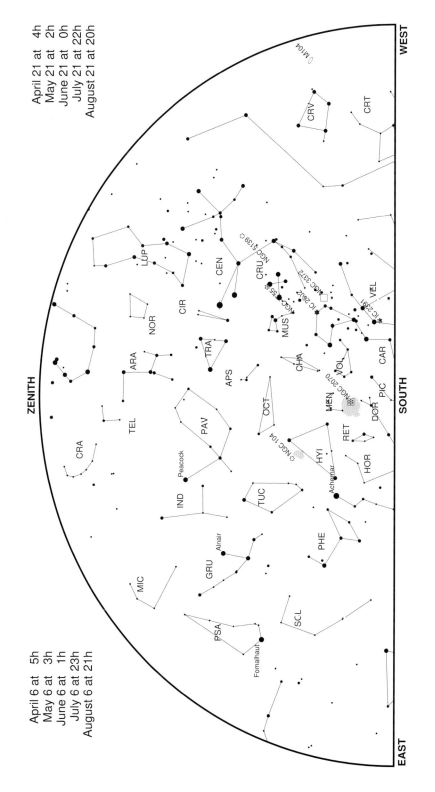

7S

WEST

EAST

ZENITH

SOUTH

April 21 at 4h
May 21 at 2h
June 21 at 0h
July 21 at 22h
August 21 at 20h

April 6 at 5h
May 6 at 3h
June 6 at 1h
July 6 at 23h
August 6 at 21h

CRV
CRT
M104

LUP
CEN
CIR
NOR
ARA
TRA
APS
CRU
NGC 5139
MUS
NGC 4755
IC 2602
NGC 3372
VEL
IC 2391
CHA
VOL
CAR
MEN
NGC 2070
DOR
PIC
RET
HOR
HYI
Achernar
NGC 104
OCT
PAV
TEL
CRA
IND
Peacock
TUC
PHE
SCL
GRU
Alnair
MIC
PSA
Fomalhaut

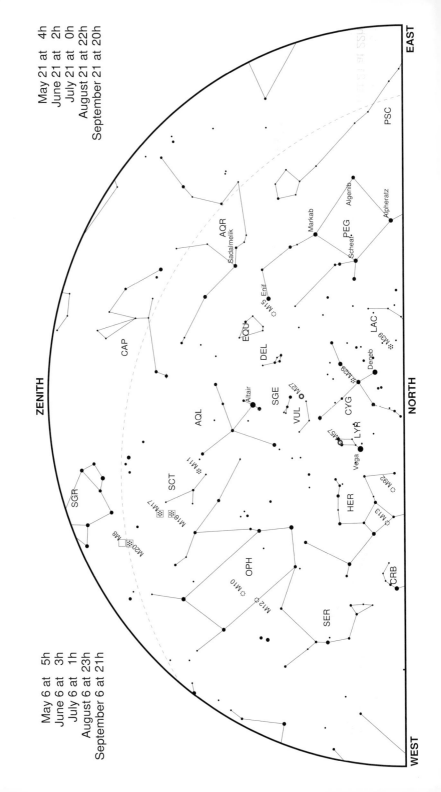

8N

May 21 at 4h
June 21 at 2h
July 21 at 0h
August 21 at 22h
September 21 at 20h

EAST

PSC

Algenib

Alpheratz

Markab

PEG

AQR

Scheat

Sadalmelik

Enif

M15

EQU

LAC

DEL

M39

ZENITH

Deneb

M29

CAP

Altair

SGE

CYG

AQL

VUL

M27

LYR

M57

Vega

SCT

M11

M92

M17

HER

M16

M13

SGR

M20

OPH

M10

CRB

M8

M12

SER

NORTH

WEST

May 6 at 5h
June 6 at 3h
July 6 at 1h
August 6 at 23h
September 6 at 21h

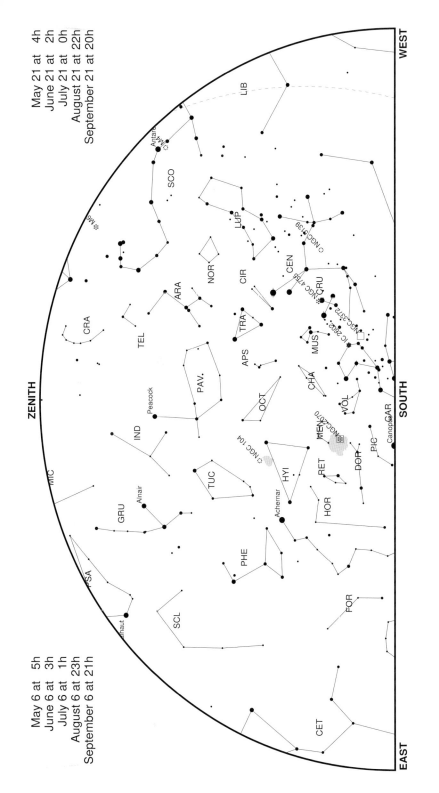

8S

WEST

May 21 at 4h
June 21 at 2h
July 21 at 0h
August 21 at 22h
September 21 at 20h

ZENITH

May 6 at 5h
June 6 at 3h
July 6 at 1h
August 6 at 23h
September 6 at 21h

EAST

SOUTH

9N

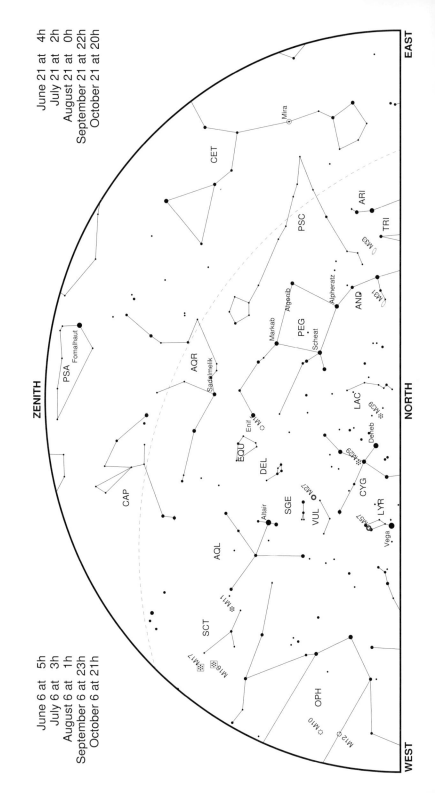

June 21 at 4h
July 21 at 2h
August 21 at 0h
September 21 at 22h
October 21 at 20h

June 6 at 5h
July 6 at 3h
August 6 at 1h
September 6 at 23h
October 6 at 21h

EAST

ZENITH

NORTH

WEST

Mira

CET

ARI

TRI

M33

PSC

Algenib

Alpheratz

AND

M31

Markab

PEG

Scheat

LAC

M39

PSA

Fomalhaut

AQR

Sadalmelik

Enif

M15

EQU

DEL

Deneb

M29

CYG

CAP

Altair

SGE

VUL

M27

LYR

M57

Vega

AQL

M11

SCT

M17

M16

OPH

M10

M12

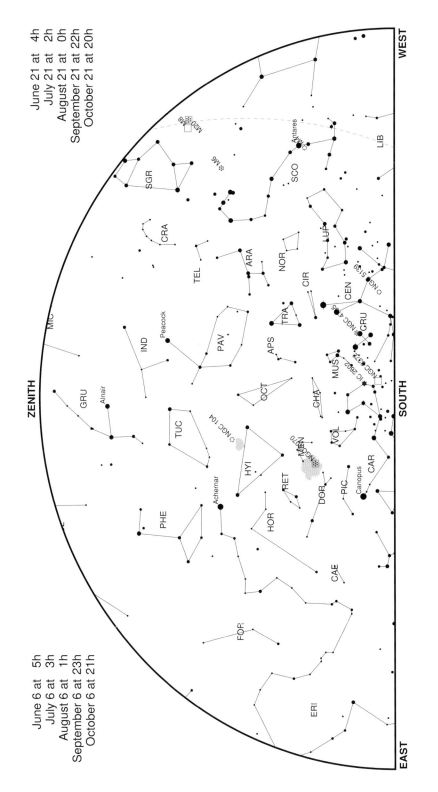

9S

WEST

June 21 at 4h
July 21 at 2h
August 21 at 0h
September 21 at 22h
October 21 at 20h

ZENITH

June 6 at 5h
July 6 at 3h
August 6 at 1h
September 6 at 23h
October 6 at 21h

EAST

SOUTH

M20
M8
M6
Antares
SCO
LIB
SGR
CRA
TEL
ARA
NOR
CIR
LUP
NGC 5139
CEN
NGC 4755
GRU
NGC 4372
IC 2602
MUS
CHA
VOL
MIC
Peacock
IND
PAV
APS
TRA
OCT
GRU
Alnair
TUC
NGC 104
HYI
RET
NGC 2070
DOR
PIC
CAR
Canopus
PHE
Achernar
HOR
CAE
FOR
ERI

10N

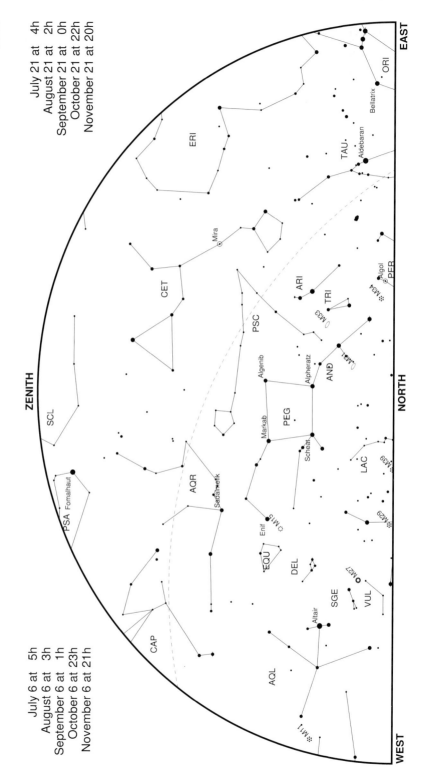

July 6 at 5h
August 6 at 3h
September 6 at 1h
October 6 at 23h
November 6 at 21h

EAST

ORI

Bellatrix

TAU

Aldebaran

ERI

Mira

ARI

TRI

M33

PER

Algol

M34

PSC

ZENITH

CET

Algenib

Alpheratz

AND

M31

SCL

Markab

PEG

Scheat

LAC

NORTH

M39

AQR

Sadalmelik

PSA

Fomalhaut

Enif

M15

EQU

DEL

M27

VUL

SGE

CAP

Altair

AQL

M71

WEST

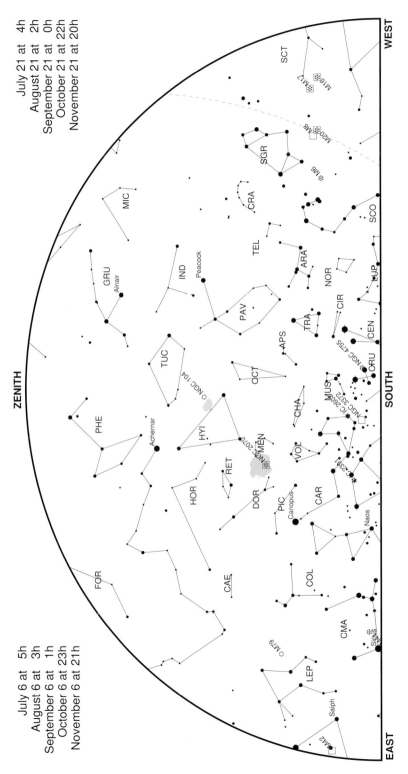

10S

WEST

July 21 at 4h
August 21 at 2h
September 21 at 0h
October 21 at 22h
November 21 at 20h

ZENITH

July 6 at 5h
August 6 at 3h
September 6 at 1h
October 6 at 23h
November 6 at 21h

EAST

SOUTH

11N

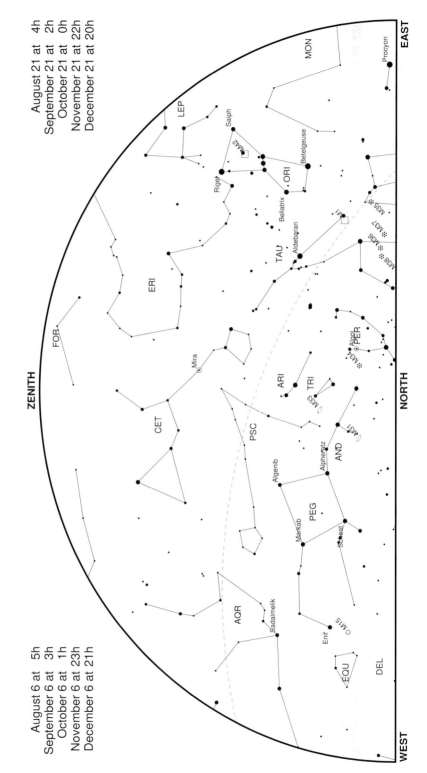

August 21 at 4h
September 21 at 2h
October 21 at 0h
November 21 at 22h
December 21 at 20h

August 6 at 5h
September 6 at 3h
October 6 at 1h
November 6 at 23h
December 6 at 21h

ZENITH

EAST

NORTH

WEST

MON
Procyon
LEP
Saiph
M42
Betelgeuse
ORI
Riget
Bellatrix
Aldebaran
TAU
M1
M35
M37
M36
M38
ERI
PER
M34
FOR
ARI
TRI
M43
Mira
M31
CET
AND
PSC
Alpheratz
Algenib
PEG
Markab
Scheat
AQR
Sadalmelik
M15
Enif
EQU
DEL

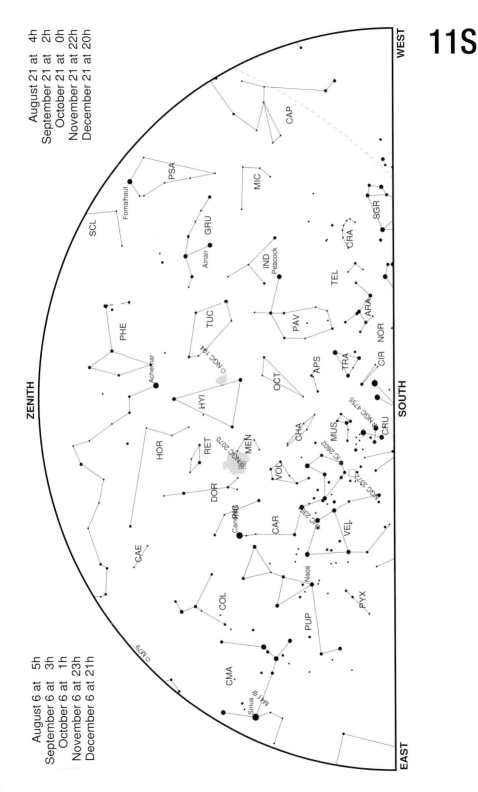

11S

WEST

August 21 at 4h
September 21 at 2h
October 21 at 0h
November 21 at 22h
December 21 at 20h

August 6 at 5h
September 6 at 3h
October 6 at 1h
November 6 at 23h
December 6 at 21h

ZENITH

EAST

SOUTH

CAP

PSA

MIC

SCL

Fomalhaut

GRU

Alnair

IND

Peacock

SGR

CRA

TEL

PAV

ARA

NOR

CIR

TRA

PHE

TUC

NGC 104

APS

OCT

Achernar

HYI

HOR

RET

MEN

CHA

NGC 2070

MUS

NGC 4755

CRU

VOL

IC 2602

NGC 3372

DOR

CAR

IC 2391

VEL

CAE

Canopus

PUP

Naos

COL

PYX

CMA

Sirius

M41

M79

12N

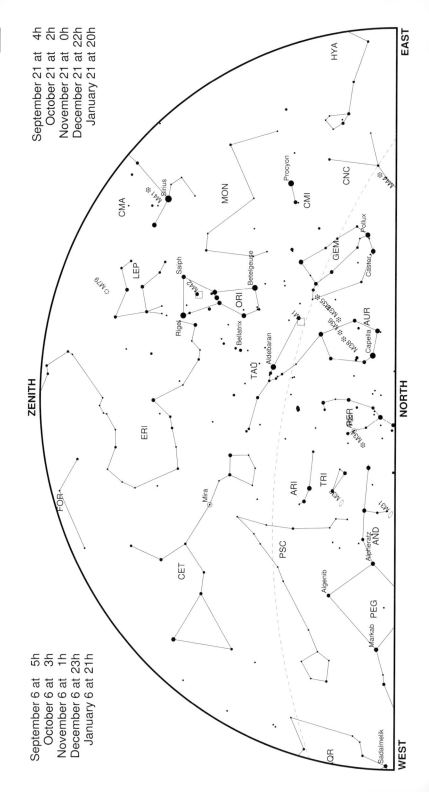

September 21 at 4h
October 21 at 2h
November 21 at 0h
December 21 at 22h
January 21 at 20h

September 6 at 5h
October 6 at 3h
November 6 at 1h
December 6 at 23h
January 6 at 21h

EAST

WEST

ZENITH

NORTH

HYA

Procyon
CMI
CNC
M44

GEM
Pollux
Castor
AUR
Capella
M37
M36
M38
M35

MON

Sirius
M41
CMA

LEP
M79

Saiph
M42
ORI
Betelgeuse
Rigel
Bellatrix

Aldebaran
TAU
LYN

PER
M34
Mirfak

ERI

TRI
ARI
M33

AND
Alpheratz
Algenib

CET
Mira
FOR

PSC

PEG
Markab

AQR
Sadalmelik

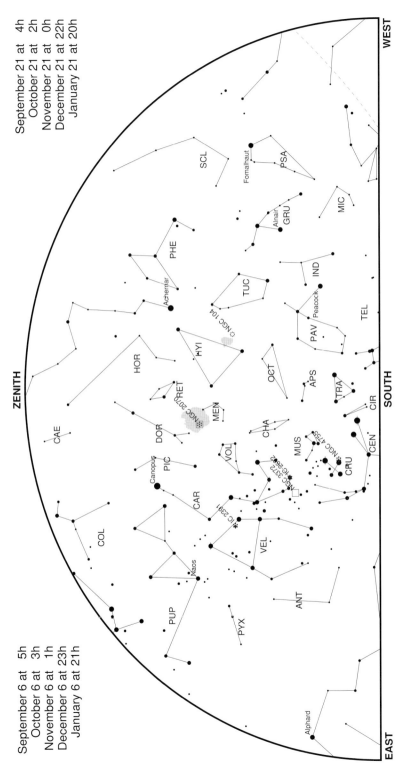

12S

September 21 at 4h
October 21 at 2h
November 21 at 0h
December 21 at 22h
January 21 at 20h

September 6 at 5h
October 6 at 3h
November 6 at 1h
December 6 at 23h
January 6 at 21h

WEST

EAST

ZENITH

SOUTH

SCL

PSA
Fomalhaut

GRU
Alnair

MIC

PHE

TUC
IND

Achernar

NGC 104

HYI

PAV
Peacock

TEL

HOR

OCT

APS

RET

MEN
NGC 2070

DOR

CHA

TRA

CAE

CIR

VOL

MUS
NGC 4755

CEN

PIC

NGC 3372
IC 2602

CRU

Canopus

IC 2391

CAR

VEL

COL

Naos

ANT

PUP

PYX

Alphard

The Planets in 2025

Lynne Marie Stockman

Mercury inhabits the twilight zone, never far from the Sun, and is often hard to spot low to the horizon in still-bright skies. Evening apparitions occur between superior conjunction and inferior conjunction, with the tiny planet starting out fully illuminated and at its maximum brightness, and ending as a narrow, sixth-magnitude crescent. Morning apparitions begin at inferior conjunction and are characterised by Mercury getting brighter every day until superior conjunction. In 2025, Mercury begins and ends the year in the morning, with three evening apparitions alternating with two further morning apparitions. For observers in northern latitudes, the best morning appearance of this tiny world is from August through mid-September; the best evening apparition is February to March. Astronomers in the southern hemisphere have their best chance of seeing Mercury in the morning from late March through May; the best evening apparitions are June–July and (even better) mid-September to mid-November. Mercury passes by all of the planets this year and it is occulted by the Moon once, in March, in an event visible after sunset from parts of the southern Pacific Ocean.

Apparition diagrams showing the position of Mercury above the eastern and western horizons may be found throughout the *Monthly Sky Notes*.

Venus rules the west after sunset as the year gets underway, with greatest elongation east taking place early in January. The evening star vanishes in late March as Venus undergoes inferior conjunction; it will spend the rest of the year in dawn skies. Greatest elongation west occurs at the beginning of June. Venus encounters all of the planets except Mars, meeting up with Mercury once in November, passing Jupiter in August, zigzagging past Saturn and Neptune three times, and outshining Uranus in July. The Moon occults Venus once, in September.

Apparition diagrams showing the position of Venus above the western and eastern horizons may be found in the January and April *Sky Notes* respectively.

The **Moon** spends the year busily occulting the first-magnitude stars Antares, Regulus (from July), and Spica (through November) as well as the Pleiades open star cluster in Taurus. Of the planets, Mercury and Venus are occulted once, Saturn and Neptune twice, and Mars four times. The minimum number of eclipses, four, occurs this year, with two in March and two in September. Both solar eclipses are partial and both lunar eclipses are total. For more information, see *Eclipses in 2025*. The Moon also reaches a major lunar standstill in March.

Mars begins the year at opposition and spends the rest of 2025 as an evening sky object. It becomes very difficult to view in the last month or two of the year as it closes in on conjunction with the Sun in January 2026. The red planet is occulted by the Moon four times although one of those instances is only visible from the Antarctic. Of all the planets only Mercury comes to call, appearing close to Mars in October and November. Starting in retrograde in the constellation of Cancer, Mars moves through the zodiac – Gemini, back to Cancer, Leo, Virgo, Libra, Scorpius – before sojourning through Ophiuchus and then on to Sagittarius by year's end. May finds the planet in the vicinity of M44, the Beehive Cluster, and in June, Mars flies past the first-magnitude star Regulus.

Finder charts showing the position of Mars around opposition and throughout the first nine months of 2025 may be found at the end of this article.

Jupiter was at opposition in December 2024 and will not come to opposition again until January 2026. It is well-place for viewing in the evening at the outset of the year, retrograding through Taurus before returning to direct motion in February. It enters Gemini in June, shortly before its conjunction with the Sun. Reappearing in the dawn sky in July, Jupiter has a close encounter with the morning star in August and moves past the planetary nebula C39 twice, once in September and again in December (when the planet is again in retrograde). The only other planet that Jupiter encounters this year is Mercury but this event takes place only days before solar conjunction. The Moon stays well away all year, never coming closer than 3.7° from the gas giant.

A finder chart showing the position of Jupiter throughout 2025 may be found at the end of this article.

Saturn undergoes both geocentric and heliocentric ring-plane crossings this year but only the latter will be visible from Earth. An in-depth look at this unusual event appears after the May *Sky Notes*. Saturn opens the year with two lunar occultations, followed by conjunction with the Sun in March. Mercury and Venus are in close attendance either side of solar conjunction. Saturn returns to the evening sky around June/July and reaches opposition on September. Two legs of a triple conjunction in right ascension with Neptune occur this year, with the final meeting occurring in February 2026. The ringed planet opens the year in Aquarius before moving into Pisces in April. Retrograde motion returns the gas giant to Aquarius in late September where it remains for the rest of the year.

A finder chart showing the position of Saturn throughout 2025 may be found at the end of this article.

Uranus begins the year in Aries, moving into Taurus in March where it will remain until 2032. It is a quiet year for this distant world, with Uranus passing by no bright stars and with the Moon never approaching closer than 4.3° to the north. The sixth-magnitude stars 13 Tauri and 14 Tauri, both of a similar brightness to Uranus, are nearby in May/June and again in December. Conjunction occurs in May, with Mercury close by, and Venus is 2.4° away from the sixth-magnitude planet in June. Uranus brightens to magnitude +5.6 during its late-November opposition.

A finder chart showing the position of Uranus throughout 2025 may be found at the end of this article.

Neptune spends the entirety of 2025 in the constellation of Pisces. It is an evening sky object at the beginning of the year, undergoing conjunction with the Sun in March and emerging into dawn skies afterwards. Neptune begins rising before midnight around June, finally reaching opposition in late September. The blue ice giant undergoes lunar occultations in January and February, and is in conjunction (in right ascension) with Saturn twice, in June and August. The third part of this triple conjunction in right ascension takes place next year, along with a single conjunction in ecliptic longitude. It is interesting to note that when Neptune was discovered in September 1846, it was in the middle of a triple conjunction with Saturn!

A finder chart showing the position of Neptune throughout 2025 may be found at the end of this article.

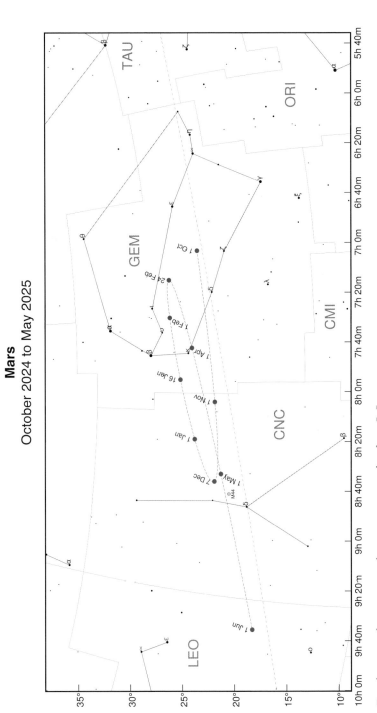

Mars
October 2024 to May 2025

Background stars are shown to magnitude +5.5.

Mars
January to September 2025

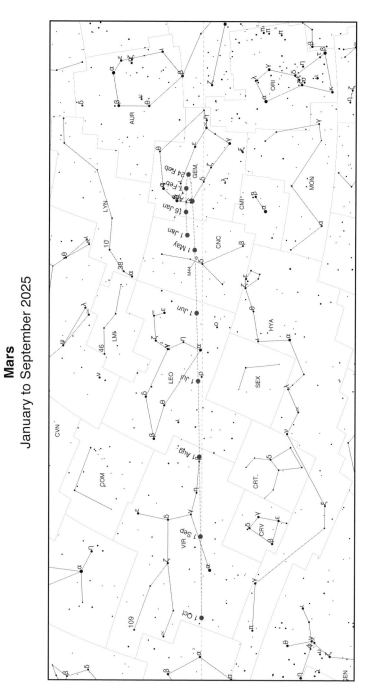

Background stars are shown to magnitude +5.5.

Jupiter
January to December 2025

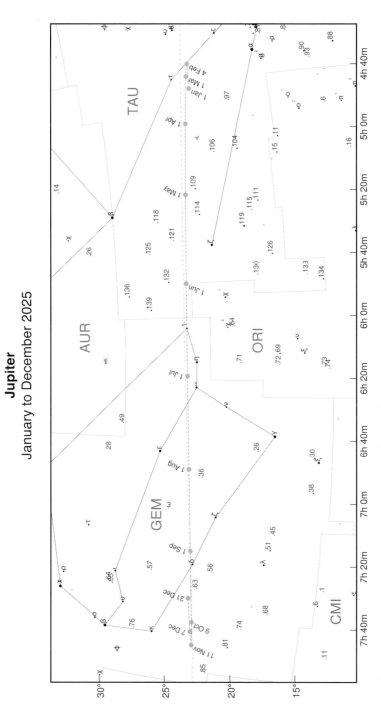

Background stars are shown to magnitude +5.5.

Saturn

January to December 2025

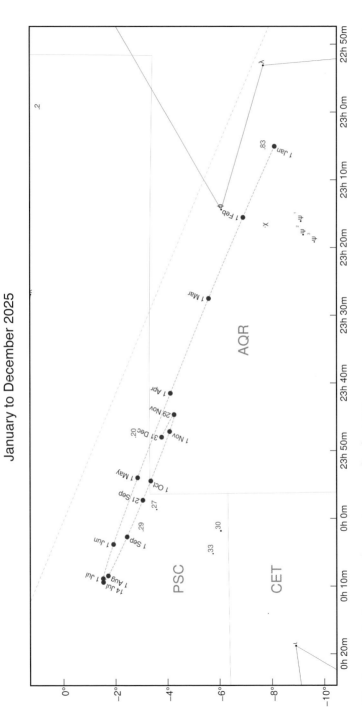

Background stars are shown to magnitude +5.5.

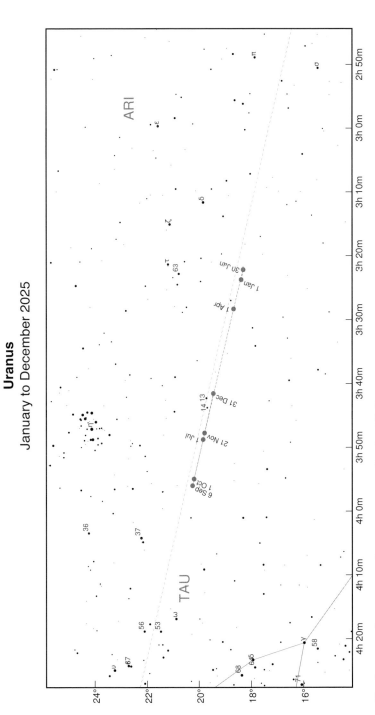

Uranus

January to December 2025

Background stars are shown to magnitude +8.0.

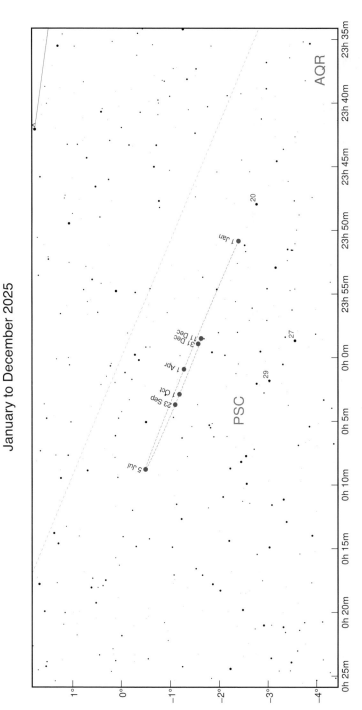

Neptune
January to December 2025

Background stars are shown to magnitude +10.0.

Lunar Phenomena and Eclipses in 2025

The dates and circumstances of all lunar phenomena are calculated from the JPL DE406 solar system ephemeris. All dates are given in UT. Positions are geocentric apparent places, referred to the true equator and equinox of date.

Phases of the Moon

The apparent shape of the Moon is ever changing; this is called its *phase*. A *lunation* or lunar month typically begins with the unlit New phase. The Moon's illuminated portion becomes larger, progressing through the waxing crescent, first quarter, and waxing gibbous phases before becoming fully lit. After Full Moon, the phase decreases, going through waning gibbous, last quarter, and waning crescent, before returning to New and the beginning of the next lunation. This cycle of phases is called a *synodic month* and its average length is 29.5 days.

New Moon	First Quarter	Full Moon	Last Quarter
	6 January	13 January	21 January
29 January	5 February	12 February	20 February
28 February	6 March	14 March	22 March
29 March	5 April	13 April	21 April
27 April	4 May	12 May	20 May
27 May	3 June	11 June	18 June
25 June	2 July	10 July	18 July
24 July	1 August	9 August	16 August
23 August	31 August	7 September	14 September
21 September	29 September	7 October	13 October
21 October	29 October	5 November	12 November
20 November	28 November	4 December	11 December
20 December	27 December		

The dates given in the above table are based on Universal Time (UT).

Apsides

An *apsis* (plural: apsides) is either the nearest or the farthest point in an elliptical orbit of one body about another. *Periapsis* is the nearest point and *apoapsis* is the farthest point. In the case of Earth, the term *perigee* is used to describe the point where a body (in this case, the Moon) is closest to Earth and *apogee* is where a body is most distant from Earth. The difference between lunar perigee and apogee distances amounts to several tens of thousands of kilometres. The line of apsides of the Moon is not fixed in space but moves approximately 41° eastwards each year, making one full circuit around the Earth in 8.85 years. In contrast, the apsidal precession of Earth around the Sun takes about 112,000 years. The time it takes the Moon to return to the same apsis is just over 27.5 days and is called an *anomalistic month*.

Perigee	Distance (km)	Apogee	Distance (km)
8 January	370,171	21 January	404,298
2 February	367,457	18 February	404,882
1 March	361,964	17 March	405,754
30 March	358,128	13 April	406,295
27 April	357,119	11 May	406,244
26 May	359,022	7 June	405,553
23 June	363,178	5 July	404,627
20 July	368,041	**1 August**	404,161
14 August	369,288	29 August	404,548
10 September	364,777	26 September	405,548
8 October	359,819	23 October	406,444
5 November	356,833	**20 November**	406,691
4 December	356,963	17 December	406,322

Distances are measured from Earth's centre to Moon's centre. Dates of extreme values (annual maxima/minima) are marked in **bold**.

Nodes

The orbit of the Moon is inclined with respect to the ecliptic by just over 5° and a lunar *node* is a point in the Moon's orbit where it intersects the ecliptic plane. The *ascending node* is where the Moon crosses the ecliptic south to north and the *descending node* is where the Moon crosses the ecliptic north to south. The time it takes for the Moon to pass through the same node twice in succession is called a *draconic month*; this is just over 27 days long. These nodes are not fixed in space but precess westward by around 19° each year, making one full circuit every 18.6 years. Solar and lunar eclipses only take place when a New or Full Moon occurs very near to a node. For details on this year's eclipses, see the article *Eclipses in 2025*.

Ascending		Descending	
5 January		19 January	
1 February		15 February	
1 March		**14 March**	Full Moon occurs 6.9 hours before node crossing: **total lunar eclipse** (**14 March**)
28 March	New Moon occurs 18.5 hours after node crossing: **partial solar eclipse** (**29 March**)	10 April	
25 April		7 May	
22 May		4 June	
18 June		1 July	
15 July		28 July	
11 August		24 August	
7 September	Full Moon occurs 5.0 hours before node crossing: **total lunar eclipse** (**7 September**)	**20 September**	New Moon occurs 20.7 hours after node crossing: **partial solar eclipse** (**21 September**)
5 October		18 October	
1 November		14 November	
28 November		11 December	
25 December			

Dates of nodal crossings resulting in eclipses are marked in **bold**.

Lunistices or Lunar Standstills

We are familiar with the term *solstice*. The word is derived from the Latin *solstitium* which means "Sun stands still". Solstices occur twice a year and mark when the Sun reaches its farthest point north or south of the celestial equator. In other words, a solstice occurs when the Sun is at its most extreme declination. On Earth, the June solstice marks the Sun's most northerly excursion from the celestial equator; astronomical summer, with its long days and short nights, begins in the northern hemisphere whilst the reverse happens in the southern hemisphere. Conversely, the December solstice is when the Sun is at its most southerly extent, bringing summer to the southern hemisphere and winter's short days and long nights to the north.

The Moon is always close to the ecliptic, so its declination undergoes a similar range of variation to the Sun, but over the course of a lunar month rather than a year. The term *lunistice* (the Latin for moon is *luna*) has been used since at least the eighteenth century to identify the most northerly and southerly declinations attained by the Moon over a draconic month. This variation in declination is most noticeable when the Moon is full. The Full Moons of mid-winter stand high above the horizon, like the mid-summer Sun, whilst the summer Full Moons are low in the sky, like the Sun in mid-winter. However, because the Moon's orbit is tilted by about 5° relative to the ecliptic and the nodes move backwards along the ecliptic in an 18.6-year cycle, the Moon's declination has a greater range than that of the Sun.

When the ascending node of the Moon's orbit is close to the First Point of Aries, where the ecliptic crosses the celestial equator from south to north, then the Moon's orbit is tilted by more than 28.5° relative to the celestial equator. For several months, the Moon's (geocentric) declination ranges from −28° to +28° every lunar month. At its extreme, the Moon's declination can be 28° 44′ north or south. Conversely, nine years later, the Moon's ascending node is close to the point where the ecliptic crosses the celestial equator from north to south, in Virgo. Then, the Moon's orbit is tilted by only 18° relative to the celestial equator.

The most noticeable effect is upon the rising and setting points of the Moon, which are further north or south on the observer's horizon than at any other time during the 18.6-year cycle. The archaeoastronomer Alexander Thom speculated that some ancient stone circles may have been built to record these extremes.

Major Lunar Standstill
Extremes of Azimuth at Moonrise

The blue disks show the extremes of the Moon's position on the eastern horizon at moonrise over three centuries (1800 to 2099), at the latitudes of Orkney, London and Sydney. For comparison, the yellow disks mark the position of the Sun at midwinter and midsummer sunrise. The horizontal scale covers 180 degrees in azimuth. Note that the blue and yellow disks are not to scale. The true size of the Moon (and Sun) is shown by the black dot at bottom left. (David Harper)

Thom coined the term *lunar standstill* in his book *Megalithic Lunar Observatories* (Oxford University Press, 1970) to describe these extreme values.

Over a period of 18.6 years, the largest lunistice is called a *major lunar standstill*. The smallest lunistice is a *minor lunar standstill*. At a major lunar standstill, the Moon's range of declination is at its greatest (over 56°) over the lunar month, as is the range of the Moon's azimuth at moonrise and moonset. Similarly, at a minor lunar standstill, both the Moon's extreme declinations (±18°) and the range of the Moon's azimuth at moonrise and moonset are at their minimum.

The lunar standstills for 2025 are given below. Declinations are geocentric.

Southerly Lunistice		Northerly Lunistice	
Date	Declination	Date	Declination
		12 January	+28° 28′
26 January	−28° 33′	8 February	+28° 36′
22 February	−28° 41′	7 March	+28° 43′
22 March	−28° 44′	3 April	+28° 42′
18 April	−28° 38′	1 May	+28° 35′
15 May	−28° 29′	28 May	+28° 27′
11 June	−28° 24′	25 June	+28° 24′
9 July	−28° 26′	22 July	+28° 29′
5 August	−28° 32′	18 August	+28° 35′
1 September	−28° 38′	14 September	+28° 38′
29 September	−28° 36′	12 October	+28° 33′
26 October	−28° 28′	8 November	+28° 23′
22 November	−28° 18′	5 December	+28° 16′
19 December	−28° 14′		

The date of the major lunar standstill is marked in **bold**.

On 22 March 2025, the Moon will reach a major lunar standstill with a declination of −28° 44′. (The most northerly declination occurs 15 days earlier, on 7 March, when the Moon reaches a declination of +28° 43′.) The next minor lunar standstill will occur on 26 March 2034 when the Moon reaches a maximum distance of 18° 09′ from the celestial equator. Following that will be a major lunar standstill on 25 September 2043 with the Moon's geocentric declination reaching +28° 43′.

Lunar Occultations

A lunar occultation occurs when the Moon passes in front of a more distant celestial body as seen from Earth, partially or totally obscuring this object and blocking its light. Like total solar eclipses, occultations are seen only at particular times and from particular places, but they have been noticed for centuries; the first known observation dates to 357 BCE when Aristotle spotted Mars disappearing behind the Moon's disk. Occultation data can help astronomers measure the heights and depths of lunar features and improve knowledge about the lunar orbit; such information can be used to detect close companions in multiple star systems and improve the precision of stellar positions. Lunar occultation data was even employed to pinpoint the location of 3C 273, the first quasar to be identified by astronomers.

The Moon occults countless faint objects every month but it also moves in front of the occasional planet or bright star, including four first-magnitude stars: α Leonis (Regulus), α Scorpii (Antares), α Tauri (Aldebaran), and α Virginis (Spica). The latest occultation series of Antares began in August 2023 and will continue through August 2028. Spica has been undergoing lunar occultations since June last year; this series will conclude in November. An occultation series of Regulus begins in July and will continue into December next year. The open star cluster M45 (Pleiades) or rather, the third-magnitude star η Tauri (Alcyone), has been regularly occulted by the Moon since September 2023 and will continue to vanish behind the Moon's disk until July 2029. For the time being, only Aldebaran remains safely out of reach of the Moon.

Both inferior planets are occulted by the Moon once this year, Mercury in March and Venus in September. Saturn and Neptune, near each other in the sky, both undergo lunar occultations in January and February. Mars is a repeat target this year, with excellent viewing opportunities in January and February, but the red planet is much nearer to the Sun for the two occultations taking place in June and July.

Table of Lunar Occultations in 2025

	Jan	Feb	Mar	Apr	May	Jun	Jul	Aug	Sep	Oct	Nov	Dec
α Scorpii (Antares)	25	21	20	16	14	10	7	4, 31	27	25	21	18
α Leonis (Regulus)							26	23	19	16	12	10
α Virginis (Spica)	21	17	16	13	10	6	3, 31	27	23	21	17	
M45 (Pleiades)	10	6	5	1, 29	26	23	20	16	12	10	6	4, 31
Mercury			1									
Venus									19			
Mars	14	9				30	28					
Saturn	4	1										
Neptune	5	1										

The dates given in the above table are based on Universal Time (UT).

Details of individual occultations listed in the above table can be found in the corresponding sections of the *Monthly Sky Notes*.

For more information on occultations, including observing predictions for this year, visit the International Occultation Timing Association (IOTA) online at **occultations.org**

Eclipses in 2025

There are a minimum of four eclipses in any one calendar year, comprising two solar eclipses and two lunar eclipses. Most years have only four although it is possible to have five, six or even seven eclipses during the course of a year. There are a total of four eclipses in 2025, comprising two solar eclipses and two lunar eclipses. It is important to note that the times quoted for each event below refer to the start, maximum and ending of the eclipse on a global scale rather than with reference to specific locations. As far as lunar eclipses are concerned, these events are visible from all locations that happen to be on the night side of the Earth. Depending on the exact location of the observer, the entire eclipse sequence may be visible, although for some the Moon will be either rising or setting while the eclipse is going on.

The first eclipse of the year is the total lunar eclipse of 14 March, which will be visible in whole or in part throughout Europe, eastern Asia, most of Australia and Africa, North America, South America, the Pacific Ocean, the Atlantic Ocean, Arctic and Antarctica. The eclipse commences when the Moon enters the Earth's penumbra (partial shadow) at 03:57 UT and ends at 10:00 UT. The Moon begins to enter the Earth's umbra at 05:10 UT, with full eclipse lasting from 06:26 UT to 07:31 UT and maximum eclipse occurring at 06:59 UT. The Moon leaves the umbra at 08:48 UT.

The partial solar eclipse of 29 March will be visible throughout most of Europe, the northern Atlantic Ocean including Iceland and Greenland, Arctic, northern Asia, north-western regions of Africa, eastern regions of North America and Canada, including Baffin Island, Bermuda and the region of Guyana, Suriname and French Guiana in South America, with the best views being available from Canada and Baffin island where the coverage will be almost 90%. The eclipse begins at 08:51 UT and ends at 12:44 UT with maximum eclipse occurring at 10:47 UT.

On 7 September there will be a total lunar eclipse, the whole or parts of which will be visible from Europe, Asia, Australia, Africa, Alaska, the eastern tip of South America, the western Pacific Ocean, Australia and New Zealand, the eastern Atlantic Ocean, the Indian Ocean, Arctic and Antarctica. The eclipse commences when the Moon enters the Earth's penumbra (partial shadow) at 15:28 UT and ends at 20:55 UT. The Moon begins to enter the Earth's umbra at 16:27 UT, with full eclipse lasting from 17:31 UT to 18:53 UT and maximum eclipse occurring at 18:12 UT. The Moon leaves the umbra at 19:57 UT.

There will be a partial solar eclipse on 21 September which will be visible from southern Australia, New Zealand and the southern Pacific Ocean, including New Caledonia, Fiji, Norfolk Island, Vanuatu, Cook Islands, Samoa, Tonga and Niue, and Antarctica, with the best views being from New Zealand. The eclipse begins at 17:30 UT and ends at 21:54 UT with maximum eclipse occurring at 19:42 UT.

Monthly Sky Notes and Articles

Evening Apparition of Venus
June 2024 to March 2025

52° North
35° South

1 Feb
9 Jan (GE east)
1 Jan
1 Mar
1 Dec
1 Nov
1 Oct
1 Jan
9 Jan (GE east)
1 Sep
1 Feb
1 Mar
1 Aug
1 Aug
1 Jul
1 Jul
1 Sep
1 Oct
1 Nov
1 Dec

50°
40°
30°
20°
10°
0°

S
SW
W
NW

January

3/4	Earth	Quadrantid meteor shower (ZHR 120)	
4	Earth	Perihelion	
4	Moon, Saturn	0.6° apart: lunar occultation	60°
5	Moon, Neptune	1.0° apart: lunar occultation	72°
6	Moon	First Quarter	90°
10	Moon, M45 (Pleiades)	0.3° apart: lunar occultation	130°
10	Venus	Greatest elongation east	47°
13	Moon	Full	175°
14	Moon, Mars	0.2° apart: lunar occultation	175°
16	Mars	Opposition	176°
21	Moon, α Virginis (Spica)	0.1° apart: lunar occultation	97°
21	Moon	Last Quarter	90°
25	Moon, α Scorpii (Antares)	0.3° apart: lunar occultation	55°
29	Moon	New	4°

Dates are based on UT. Lunar occultations of M45 refer to η Tauri (Alcyone). Peak activity dates for meteor showers are estimates. The last column gives the approximate elongation from the Sun.

Mercury is found in the east at dawn. It is best seen from equatorial and southern latitudes but is already heading back toward the horizon. Mercury is magnitude −0.4 at the beginning of year but brightens to −1.0 as it gets lower in the sky. The tiny planet passes through its descending node relative to the ecliptic on 9 January and reaches its first aphelion of 2025 ten days later.

Venus is the evening star, known to the Romans as *Vesper*. At sunset it is found in the southwest for observers in the northern hemisphere whereas it is much closer to due west for astronomers located south of the equator. As the month progresses, Venus continues to gain altitude for those looking for it from northern temperate latitudes but it is already descending back toward the horizon for planet watchers in the southern hemisphere. Venus, shining at magnitude −4.4, makes a pretty picture with the crescent Moon on the third day of the month when the two bodies are just 1.3° apart. Greatest elongation east takes place on 10 January when the bright planet is 47.2° from the Sun. Theoretical dichotomy (when Venus is half illuminated as seen from Earth) is two days later. Venus passes through its

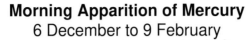

Morning Apparition of Mercury
6 December to 9 February

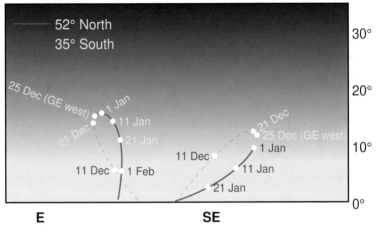

ascending node on 16 January; it will remain north of the ecliptic until early May. The bright planet is found 2.2° north (in ecliptic latitude) of Saturn on 19 January.

Earth enjoys dark skies for this year's return of the Quadrantid meteor shower which peaks in the first few days of the month. The article *Meteor Showers in 2025* has more information. The Quadrantids always occur around the same time as Earth's perihelion, which this year takes place on 4 January. The Moon has a busy month, occulting Saturn on 4 January, Neptune the following day, and Mars on 14 January. Stellar objects are also targets, with the Pleiades disappearing behind the Moon's disk on 10 January, first-magnitude star Spica occulted on 21 January, and first-magnitude star Antares vanishing on 25 January. In addition, the Moon's most distant perigee of the year occurs just after midnight on 8 January.

Mars opens the year in Cancer before retrograding into Gemini on 12 January. It is occulted by the waning gibbous Moon two days later in an event visible from much of North America and parts of western Africa. The occultation begins at around 02:00 UT. Opposition takes place on 16 January when the red planet, shining at magnitude −1.4, presents a disk 14.5″ across in a telescope. Nearest approach to Earth occurs four days earlier, on 12 January, when the two planets are 0.64 au apart. On 28 January, Mars is found 0.9° south of the fourth-magnitude M-type giant star υ Geminorum.

Jupiter was at opposition last month and is visible most of the night, setting just before dawn. The giant planet is in retrograde in Taurus, shining at magnitude −2.6 a little over 5° north of Aldebaran, the brightest star in that constellation.

Saturn is occulted by the waxing crescent Moon on the fourth day of the month. This event, which is already in progress at sunset, is visible from north-western Africa, Europe, Iceland, and western Russia. The rings of the planet are at their most open at the beginning of the year; they will close down to edge-on in March. Venus and Saturn are just over 2° apart in western skies on 19 January but first-magnitude Saturn is much fainter than the brilliant evening star. Look for Saturn in the constellation of Aquarius soon after the Sun sets.

Uranus returns to direct or prograde motion near the end of the month after entering the year in reverse. Located in Aries, the sixth-magnitude ice giant sets well after midnight for those staying up late in northern temperate latitudes. It is strictly an evening sky object for observers in the southern hemisphere.

Neptune is visible in the west after dark, best seen from northern latitudes where it sets well after sunset. The waxing crescent Moon occults the eighth-magnitude object on 5 January; this event begins around 14:30 UT and is visible from Greenland, Scandinavia and north-eastern Europe. However, this faint planet is best sought (with a telescope) on a moonless night during the latter part of the month. It is in the constellation of Pisces.

The Speed of Light

Neil Haggath

The velocity of light in vacuum, denoted by c, is one of the most important and fundamental constants in physics and astronomy. It's the "speed limit of the Universe", which nothing can exceed, and the fact that it *is* a constant forms the basis of Einstein's Special Theory of Relativity. Its value is now known to fantastic accuracy – better than one part in a billion – thanks to modern electronic methods. But some may be surprised to learn that it was first measured 350 years ago. How it was measured back then, without any sophisticated modern instruments, is remarkably simple, at least in theory, and required no more than a good telescope and an accurate clock.

In those days, no-one had any idea what the speed of light was, except that it was

Ole Rømer. (John McCue)

extremely fast. Galileo and others had initially tried to measure it in the same way as they had measured the speed of sound, but had soon realised that it was far too great to measure by any method involving distances on Earth. Any method of determining it would have to involve somehow measuring the travel time of light over astronomical distances.

In 1675, the Danish astronomer Ole Rømer (1644–1710), then working at the Paris Observatory under Giovanni Cassini, realised how it could be done, using observations of the satellites of Jupiter. He arrived at a value within about 7% of the true value, which was pretty good for the first attempt!

The four Galilean satellites had been known for decades, and their orbital periods had been measured pretty accurately. We are all familiar with the way the satellites and their shadows periodically transit, or cross in front of, Jupiter itself from our point of view; most of us who own telescopes have observed such transits for ourselves. They were easy to observe, even with the telescopes of 1675.

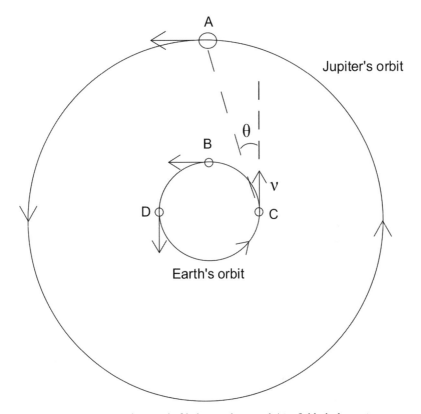

How to measure the speed of light. (Neil Haggath / Garfield Blackmore)

Some years earlier, it had been proposed to use accurate timings of these phenomena as a method of determining longitude, although it proved impractical to make the necessary observations at sea. Both Cassini and Rømer had made many observations in France and Denmark respectively, for that purpose, but found discrepancies between the predicted and observed timings – when observing from different points in the Earth's orbit – of up to ten minutes. Rømer reasoned that the discrepancies were due to the finite speed of light; he then realised that they were the key to measuring it. He presented a paper to the French *Académie des Sciences* in August 1676.

Rømer's method involved observing successive transits of the satellites, and accurately measuring the time intervals between them, when the Earth and Jupiter were at different relative positions in their orbits. We will consider Io, which has an orbital period of roughly 1.8 days.

The diagram shows the orbits of Earth and Jupiter, though not to scale. Each planet orbits anticlockwise, as seen from above its north pole. The Earth travels with a velocity 'v', which can easily be calculated, if you know the radius of its orbit. (Of course, v varies slightly during the year, as the orbit is elliptical, but for simplicity we will assume here that it is circular.) In 1675, the Earth's distance from the Sun wasn't known to great accuracy; it had been determined with an uncertainty of about ± 3%. (The transit of Venus expeditions were 86 years in the future). This was, in fact, the primary factor which limited the accuracy of Rømer's calculations. At any given point in its orbit, the Earth is moving in a direction tangential to the circle, as indicated by the arrows.

First, consider what happens when Jupiter is at position A and Earth at position B – i.e. when Jupiter is at opposition. Both planets are moving very nearly parallel to each other, so the distance between them doesn't change appreciably over a few days. We observe successive transits of Io, and measure the time interval between them, which we will call 't'.

Now, suppose we do the same when the Earth is at position C relative to Jupiter. This time, the measured interval between successive transits is slightly shorter than that measured at B, by an amount 'delta t', or δt; in other words, it is equal to $t - \delta t$.

This is because, during the time between the two transits, the Earth has moved slightly closer to Jupiter – so the light which reaches us from the second transit has travelled a slightly shorter distance than that from the first transit.

Naturally, when the Earth is at position D, the exact opposite happens, as the Earth is moving away from Jupiter. The measured time interval is longer than that measured at B, by the same amount – i.e. it is equal to $t + \delta t$. In the case of Io, δt is about 14 seconds. However, by observing transits separated by many orbital periods, as Rømer did, the difference adds up to several minutes.

If we measure the angle 'θ' between the direction of the Earth's tangential motion and that of the line of sight to Jupiter, then the component of its velocity in the direction towards Jupiter is ($v \cos \theta$). So in the time interval t, the Earth moves closer to Jupiter by a distance equal to ($vt \cos \theta$).

This distance is also the difference in the path lengths travelled by the light from the two transits. If we call the speed of light 'c', we can see that this difference in the light paths is equal to ($c\,\delta t$).

So we can now say that:

$$c\,\delta t = vt \cos \theta$$

Solve that equation, and we have the speed of light!

February

1	Moon, Saturn	1.0° apart: lunar occultation	35°
1	Moon, Neptune	1.3° apart: lunar occultation	45°
5	Moon	First Quarter	90°
6	Moon, M45 (Pleiades)	0.5° apart: lunar occultation	102°
9	Mercury	Superior conjunction: morning → evening	2°
9	Moon, Mars	0.8° apart: lunar occultation	147°
11	Uranus	East quadrature	90°
12	Moon	Full	177°
17	Moon, α Virginis (Spica)	0.3° apart: lunar occultation	125°
20	Moon	Last Quarter	90°
21	Moon, α Scorpii (Antares)	0.4° apart: lunar occultation	83°
28	Moon	New	2°

Dates are based on UT. Lunar occultations of M45 refer to η Tauri (Alcyone). The last column gives the approximate elongation from the Sun.

Mercury undergoes superior conjunction on 9 February, swinging around the far side of the Sun and reappearing in the evening sky. As is always the case with evening appearances, Mercury begins bright (magnitude −1.6 in this instance) and dims throughout the apparition, ending the month at a still-bright magnitude −1.0. It is 1.4° north of Saturn on 25 February but both planets are very close to the Sun. Mercury returns to the north side of the ecliptic on the penultimate day of the month when it reaches it ascending node. This is the best evening apparition of the year for observers in northern temperate latitudes but the worst for planet watchers further south.

Venus continues to dominate western skies after sunset, reaching its maximum brightness of −4.6 mid-month. For astronomers in northern temperate zones, the ascent of Venus into dusky evening skies halts and the bright planet begins to return toward the horizon. This falling out of the sky is even more rapid for observers further south. The evening star is in the vicinity of Neptune for the first few days of the month, outshining that distant world by an astonishing 12 magnitudes! Venus reaches perihelion for the first time this year on 19 February. Retrograde motion (in right ascension) begins near the end of the month, heralding inferior conjunction in March.

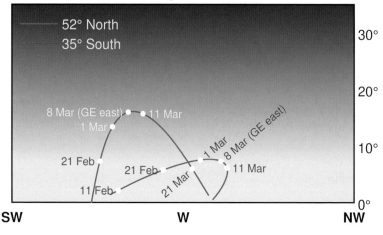

Evening Apparition of Mercury
9 February to 24 March

The **Moon** repeats last month's performance by occulting three planets, an open star cluster, and two first-magnitude stars: Saturn and Neptune on the first day of February, the Pleiades on 6 February, Mars three days later, Spica on 17 February, and finally Antares on 21 February. This is the last time the Moon will occult Saturn and Neptune this year but it is getting closer to Regulus (α Leonis), approaching to within 2° of the star on 16 February. By July, the Moon will occult Regulus on a regular basis.

Mars undergoes its second lunar occultation of the year on 9 February, with the waxing gibbous Moon passing in front of the red planet as seen from vantage points in Ellesmere Island (Canada), Greenland, Iceland, northern Scandinavia, most of Russia, eastern Kazakhstan, Mongolia, China, Japan, and North and South Korea. This event begins around 18:00 UT. Mars completes its retrogression in Gemini on 24 February and returns to direct motion which it will maintain for the rest of the year. For observers in northern temperate latitudes, Mars is visible for most of the night, not setting until dawn, but for those in the southern hemisphere, Mars vanishes just after midnight.

Jupiter returns to direct motion on 4 February. It is slowly getting fainter as Earth draws further away from it, going from magnitude −2.5 to −2.3 over the course of the month. It is found north of α Tauri (Aldebaran) and is best viewed from the northern hemisphere where the gas giant does not set until well after midnight. Observers in southern latitudes lose sight of this bright planet in late evening.

Saturn undergoes its final lunar occultation of 2025 on the first day of the month. This event is visible from north-eastern Russia after sunset. Later that same day, Saturn is 0.9° south of φ Aquarii, a fourth-magnitude spectroscopic binary system. The first-magnitude planet is getting ever lower in the west as the skies darken. Sun-skimming Mercury is a little over a degree away on 25 February by which time Saturn may be too deep into the evening twilight to be seen. The same can be said of the close pass by the very young crescent Moon on the last day of the month.

Uranus attains east quadrature on 11 February in the constellation of Aries, the Ram. This places it 90° east of the Sun in the evening sky. For southern hemisphere observers the faint planet sets mid-evening but those seeking Uranus from northern latitudes have until midnight to spot it. As the planet is never brighter than sixth magnitude, choose a dark moonless night for observations of this distant world.

Neptune also undergoes its final lunar occultation of the year on the first day of February. The waxing crescent Moon will move in front of the disk of the faint planet beginning around 22:00 UT as seen from north-eastern Russia. Bright Venus is found within 4° of Neptune during the first few days of the month. Located in Pisces, the eighth-magnitude object is most easily viewed from northern latitudes where it sets in dark skies but it is getting closer to the Sun and lower to the western horizon with every passing day.

An Astrobiological Analogy

David M. Harland

As Christopher P. McKay of the NASA Ames Research Center in California wrote in 2014,[1]

> "The primary goal of astrobiology is the search for evidence of life beyond the Earth. It is worth asking why is this search of interest and how can it be done? I submit that the answer to the first question is also the answer to the second question, and is rooted in biochemistry."

We can regard life in terms of 'hardware' consisting of a small number of molecules – namely proteins composed of amino acids, information-bearing molecules made up from five nucleotide bases, polysaccharides using a few simple sugars, and a few distinct lipids – and the 'software' of genetic information which, being common to all life on Earth, implies a common ancestor.

However, there is no consensus regarding how life started on Earth, or whether it originated elsewhere and was imported. Nor have we (yet) been able to reproduce it in the laboratory. Nevertheless, the ambition of astrobiology is to understand how biochemistry could have given rise to life, not only as we know it, but also independently elsewhere.

Many of the core biomolecules of terrestrial life – and in particular the amino acids, the so-called 'building blocks' of life – possess chiral asymmetry, meaning that they exist in two forms which are mirror images, prompting the analogy of right-handed and left-handed forms. Life, however, uses only left-handed amino acids in the construction of proteins.

Studies of the twenty or so amino acids used in proteins suggest that roughly half, the simplest, were inherited by early life from prebiotic synthesis and that the others were invented by biology.

1. 'The Search for Life on Other Worlds: Second Genesis', Christopher P. McKay, *Biochemical Society*, December 2014.

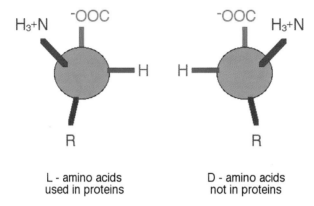

L - amino acids
used in proteins

D - amino acids
not in proteins

Amino acids have left-handed and right-handed forms, but the proteins of terrestrial life use only left-handed ones. (Courtesy of Chris McKay)

But laboratory studies have shown that entirely 'unnatural' building blocks using different amino acids can yield viable equivalents of the 'transcription machinery' of genetic molecules.

Life that originates in water is likely to use the same set of simple amino acids, both because they are of prebiotic origin and because, comprising only a small number of carbon atoms, the number of variations is small. Indeed, the simplest, glycine, has been found in meteorites and comets, and is likely to be universal.

On the other hand, large amino acids with many carbon atoms, such as histidine, are not produced in prebiotic syntheses; they must have been created by biochemical evolution.

It is here that the question of genesis becomes particularly interesting. As McKay says,

"Life using amino acids may well need to evolve structures with some of the functionality of histidine, but the chances of independently selecting the same molecule are small. When all ten complex amino acids are considered, the probability that a second genesis of life that uses amino acids in water for protein construction is based on the same twenty amino acids with the same left-handed preference is astronomically small."

McKay then presents a very insightful analogy for how a wide range of amino acids could yield carbon-based life surprisingly different to that with which we are familiar.

Consider five science books in a library, one written in English, one in Spanish, one in Russian, one in Arabic, and one Chinese, all explaining how to make a book from paper and cloth.

Although the book in Spanish shares the same alphabet as the book in English, it is clearly of a different origin. The book in Russian shares a number of the same letters, but introduces others. The book in Arabic uses an alphabet which has no symbols in common with English. The book in Chinese uses a system which has no simple relationship to the small number of alphabet characters of the other books; in fact, it uses over five thousand unique characters that represent words, ideas, and sounds.

Being made of paper and ink, all five books are the same at the bottom level, and by explaining how to make a book they are the same at the highest level. Where the books differ, is in how the ink is arranged on the paper to convey the necessary information.

This is analogous to the independent genesis of life on different worlds, perhaps in different star systems. Life would be the same at the bottom level, in that it is

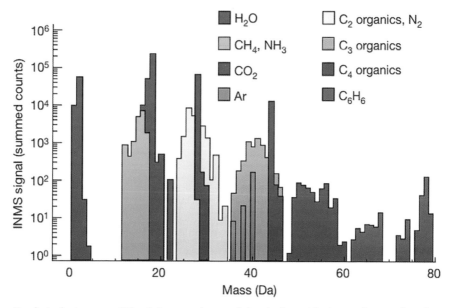

Cracks in the icy crust of Enceladus vent plumes of chemically enriched water from a subsurface ocean. This histogram shows the organic molecules detected by the Cassini spacecraft when it flew through the plumes. ('Liquid water on Enceladus from observations of ammonia and ^{40}Ar in the plume', J. H. Waite et al., *Nature*, **460**, 487–490, 2009)

made of carbon and living in water, and would be the same at the highest ecological level. Of interest here are the mid-levels of the specific molecules that implement the processes essential to life.

As McKay points out, "searching for alien biochemistry is the answer to the two questions I posed at the start."

The robotic missions we send to seek evidence of life on other worlds, perhaps long-extinct, should therefore seek 'biosignatures' in the form of molecules that are indicative of biological processes, with particular emphasis on the amino acids involved.

If we succeed, then, as McKay notes, "our scientific understanding of how carbon-based biochemistry works is likely to be greatly advanced by having two different examples to compare."

Currently the focus of the search for life in our solar system is with the Curiosity and Perseverance rovers on Mars, with the latter collecting samples that might one day be returned to Earth for laboratory study, but there are plans afoot to send one-way missions to the Jovian moon Europa and the Saturnian moon Enceladus, both of which have oceans beneath shells of ice and, in the case of Enceladus, geysers which spew chemically enriched water into space.

If it can be established that life originated independently multiple times in the solar system, that would strongly suggest the universe is teeming with life… potentially with a wide variety of biochemistries. If attempts to detect radio or laser signals from alien civilisations prove successful, one of the first things we will want to know is how their biochemistries work.

I would like to thank Chris McKay for reviewing the draft of this article.

March

1	Moon, Mercury	0.4° apart: lunar occultation	16°
2	Jupiter	East quadrature	90°
5	Moon, M45 (Pleiades)	0.7° apart: lunar occultation	75°
6	Moon	First Quarter	90°
8	Mercury	Greatest elongation east	18°
12	Saturn	Conjunction	2°
14	Moon	Full: total lunar eclipse	180°
16	Moon, α Virginis (Spica)	0.3° apart: lunar occultation	152°
19	Neptune	Conjunction	1°
20	Earth	Equinox	
20	Moon, α Scorpii (Antares)	0.5° apart: lunar occultation	110°
22	Moon	Major lunar standstill	90°
22	Moon	Last Quarter	90°
23	Venus	Inferior conjunction: evening → morning	8°
23	Saturn	Geocentric ring plane crossing	10°
24	Mercury	Inferior conjunction: evening → morning	3°
29	Moon	New: partial solar eclipse	1°

Dates are based on UT. Lunar occultations of M45 refer to η Tauri (Alcyone). The last column gives the approximate elongation from the Sun.

Mercury undergoes its only lunar occultation in 2025 on the first day of the month but the majority of this event takes place during daylight hours and is not visible. It comes to within 2° of Neptune the following day but that faint world will not be visible in the evening twilight. Mercury's first perihelion of 2025 occurs shortly thereafter, on 4 March. The closest planet to the Sun reaches a greatest elongation east of 18.2° on 8 March and makes a distant pass by Venus in the following days. This is a poor apparition for those seeking Mercury from the southern hemisphere but the best evening appearance of the year for observers much further north. Beginning the month at magnitude −1.0, the planet fades in brightness as it heads back toward the horizon, with retrograde motion beginning mid-month and inferior conjunction taking place on 24 March. When it reappears in the east at dawn, it passes by Neptune for a second time this month.

Venus is in retrograde this month. Still visible at the beginning of the month in the west at sunset, it is plummeting toward the horizon and vanishes from the evening sky before mid-March for southern astronomers. However, because Venus passes 8° north of the Sun at inferior conjunction, eagle-eyed observers in northern latitudes may be able to spot Venus as both the morning star and the evening star very low to the eastern and western horizons in the days either side of 23 March. Venus passes 5.6° north (in ecliptic latitude) of Mercury on 11 March, and as the morning star, it once again goes past Neptune (9.2° distant) on 27 March. However, the two planets are only 3° away from the Sun at the time.

Earth enjoys the first two eclipses of the year this month, with a total lunar eclipse taking place on 14 March and a partial solar eclipse occurring on 29 March. The event on 14 March is the first total lunar eclipse since November 2022. For more information on these events, see *Eclipses in 2025* elsewhere in this volume. The very young crescent Moon occults Mercury on the first day of March. This is followed by an occultation of the Pleiades star cluster on 5 March, of Spica on 16 March, and of Antares on 20 March. Our planet reaches the Vernal Equinox on 20 March, heralding spring in the northern hemisphere and introducing autumn to the southern hemisphere. Finally, on 22 March, just hours before its reaches First Quarter, the Moon undergoes a major lunar standstill, reaching an extreme southerly declination of −28° 43′ 32.4″. (The most northerly declination, slightly smaller at +28° 43′ 00.0″, is on 7 March.) Both of these values are geocentric; topocentric declinations can be a degree greater or smaller. The next major lunar standstill will not occur until 2043. For more information on lunar standstills, see the section *Lunar Phenomena: Lunistices or Lunar Standstills*.

Mars is in Gemini, becoming dimmer (starting the month at magnitude −0.3 and ending at +0.4) as Earth draws away. It is an evening sky object and is well-placed for northern observers for whom it does not set until morning twilight. For astronomers in the southern hemisphere, however, Mars vanishes before midnight. The Moon avoids occulting the red planet this month but does pass 1.6° north of Mars just after midnight UT on 9 March.

Jupiter is an evening sky object, disappearing by mid-evening for those in the southern hemisphere but remaining above the horizon until just after midnight for those inhabiting northern temperate latitudes. It reaches east quadrature on the second day of the month and two days later, is found 1.1° south of the bluish fourth-magnitude spectroscopic binary star τ Tauri. Jupiter continues to dim as it moves away from opposition and toward conjunction, ending the month at a still-brilliant magnitude −2.1.

Saturn is at conjunction with the Sun on 12 March and is lost to view this month. Its geocentric ring-plane crossing takes place only 11 days later, on 23 March, but the planet is far too close to the Sun to observe this interesting phenomenon. For more information on this year's ring plane crossing, see *Saturn At Its Equinox: The Ring Plane Crossing of 2025* following the May *Sky Notes*. The waning crescent Moon passes less than 2° north of the planet on 28 March but no lunar occultation takes place.

Uranus leaves Aries for Taurus early in the month; it will remain in the constellation of the Bull for the rest of the year. The sixth-magnitude planet sets not long after the end of astronomical twilight for southern hemisphere planet watchers, but observers in northern latitudes have until late evening to look for Uranus. Optical aids are useful for spotting this elusive gas giant as are dark skies; choose a moonless night late in the month.

Neptune is at conjunction with the Sun on 19 March and is lost to view this month. It is found near Mercury at the beginning and end of the month, and also in the vicinity of Venus days after conjunction but its proximity to the Sun (and locations in dusk and dawn skies) makes observation all but impossible.

The Smell of Space

Jonathan Powell

With the increasing ability to visually capture some of the more distant wonders of the cosmos, the true glory and majesty of what was initially a hazy image has been transformed into a spectacular portrait, encapsulating magnificent colour and detail of stunning clarity. The true testimony to these advances in technology have in recent times been chiefly attributable to the *Hubble Space Telescope* which launched in 1990, followed by the *James Webb Space Telescope*, launched in 2021.

The relaying to an eagerly waiting world of new pictures from the depths of space reinforces the optical side of astronomy, adding to the already brimming gallery. However, the same sense of 'capture' felt in recent history would have been embraced centuries ago, perhaps by colleagues of Italian astronomer Galileo Galilei (1564–1642), upon seeing his sketches of Saturn and the extraordinary 'discs' apparent on either side of the planet. A far cry perhaps from the 'Pillars of Creation', but in essence a distinct pictorial parallel across the generations.

The advances in radio astronomy pioneered by American physicist and radio engineer Karl Guthe Jansky (1905–1950) and radio enthusiast Grote Reber (1911–2002), have meant that our once muffled ear cocked toward the universe has become somewhat sensitive and as a consequence more finely tuned, from the eerie electron "whistler waves" produced by the Van Allen Belts, to NASA's capturing of the 'sound' produced by a black hole, some 250 million light-years distant.

We have a galaxy of photographs and a library full of sound, but what of the smell of space? Not to be left out, like those of sight and sound, the scent trail also follows a similar pattern, from space pioneers like the Apollo 11 astronaut Buzz Aldrin reporting a strange smell attained from lunar surroundings during the historic first visit to the Moon, to the pungent clouds of gas deep in space.

In actuality and perhaps often overlooked, space carries a whole array of odours to match the vibrancy of sights and sounds, and to start, we need not travel light-years to find an 'object' generating one such smell, the Winchcombe meteorite.

On 28 February 2021, a blazing meteorite was captured by an array of cameras across a swathe of the UK, some static, ensconced in doorbells, others mounted on vehicle dashboards, plus a multitude of cameras specifically in place to witness such an event should it occur. After the meteorite's impact in this village in

A view of the interior of Apollo 11 and Edwin E. 'Buzz' Aldrin who, along with colleagues Neil Armstrong and Michael Collins, made the landmark first flight to the Moon in July 1969. (NASA)

Gloucestershire, various fragments were duly recovered over the days that followed its fiery landing, as an excited public cautioned to not touch any remnants, busily searched Winchcombe and neighbouring territory.

Some of those who came into contact with the meteorite pieces reported a "compost-like" smell emanating from the treasure-like find. The smell gave astronomers a distinct clue to the wanderer's potential origin. The presence of organic compounds classified the Winchcombe meteorite as carbonaceous chondrite in composition, a relic from the formation of the solar system.

From the hunting of meteorites to the brave souls returning home from space, only to discover upon the removal of their suited helmets a smell akin to that of seared steak, hot metal, and welding fumes. Puzzled faces greeted the aroma; where did it come from?

One credible explanation for the smell suggested a possible chemical reaction by substances carried on the spacesuit which – during airlock re-entry after being out

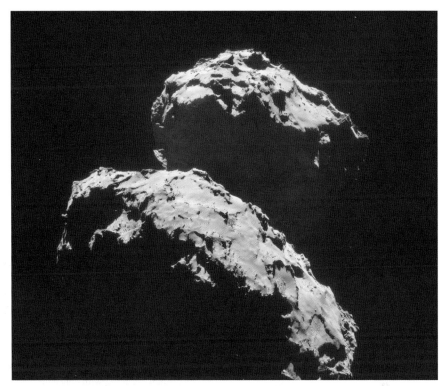

A mosaic of Comet 67P/Churyumov-Gerasimenko, the first comet to be orbited and landed upon by spacecraft from Earth. (ESA/Rosetta/NAVCAM/CC BY-SA 3.0 IGO)

in space and subsequent re-pressurization – created a sort of oxidization process. The process, like combustion, would generate a similar sort of smell, which ended up impressed on the spacesuit.

Another equally credible (but somewhat mind-blowing) explanation is that this, and indeed other space smells encountered in these situations, relate to the remnants of long since dead stars, whose essence literally fills the cosmos. Called polycyclic aromatic hydrocarbons, these are the by-products of combustion, also found in coal, oil, and even food.

When back onboard Apollo 11, astronaut Buzz Aldrin commented that his suit – soiled with lunar dust accumulated from the Moon's surface – smelled like "burnt charcoal". Harrison Schmitt onboard Apollo 17, the last of the Apollo missions to the Moon, stated that, "All I can say is that everyone's instant impression of the smell, [of the Moon], was that of spent gunpowder."

ESA's *Rosetta* mission, which rendezvoused with Comet 67/P Churyumov-Gerasimenko in 2014, was armed to the teeth with instruments, one of which was the Rosetta Spectrometer for Ion and Neutral Analysis (ROSINA) instrument. ROSINA was to inhale a toxic mix of rotten eggs (hydrogen sulphide), horse urine (ammonia), almond (hydrogen cyanide), along with formaldehyde. A potent mix, but within those smells important clues as to how our solar system may have formed.

Reaching out further into the depths of space we find Sagittarius B2, a molecular cloud of gas, 28,000 light years distant. Research has revealed the presence of an array of chemicals including ethanol, vinyl ethanol, and methanol. However, it is the amounts of ethyl formate, the chemical responsible for giving raspberries their smell, which when mixed with the other portions of the cocktail, would make Sagittarius B2 smell of raspberry rum.

Although there may well be plenty to see and hear in the cosmos, let us not forget the variety of rich and pungent smells.

A colour-composite image of the galactic centre Sagittarius B2. Measuring 150 light years across, the cloud is situated 390 light years from the centre of the Milky Way. (ESO/APEX & MSX/IPAC/NASA)

April

1	Moon, M45 (Pleiades)	0.6° apart: lunar occultation	48°
5	Moon	First Quarter	90°
13	Moon	Full: smallest apparent diameter (1766″)	178°
13	Moon, α Virginis (Spica)	0.3° apart: lunar occultation	177°
16	Moon, α Scorpii (Antares)	0.4° apart: lunar occultation	137°
17	Mercury, Neptune	0.7° apart	27°
21	Mars	East quadrature	90°
21	Moon	Last Quarter	90°
21	Mercury	Greatest elongation west	27°
22/23	Earth	Lyrid meteor shower (ZHR 18)	
27	Moon	New	3°
29	Moon, M45 (Pleiades)	0.6° apart: lunar occultation	21°

Dates are based on UT. Lunar occultations of M45 refer to η Tauri (Alcyone). Peak activity dates for meteor showers are estimates. The last column gives the approximate elongation from the Sun.

Mercury returns to direct motion in the first week of April, which brings it past Neptune for the third time this year on 16–17 April. Although the two planets are 27° away from the Sun, eighth-magnitude Neptune may be impossible to spot in dawn skies even with a telescope. Mercury reaches aphelion for the second time in 2025 on 17 April and attains greatest elongation west (27.4°) on 21 April. This dawn apparition is the best of the year for early risers in the southern hemisphere and in the regions of the tropics, with the brightening planet (magnitude +2.9 down to +0.1 by the end of the month) soaring 20° or more above the eastern horizon. However, for observers in northern temperate latitudes, this is the worst morning appearance of Mercury this year, with the tiny planet struggling to reach more than a few degrees in altitude.

Venus, in its guise as the morning star, was known to the Romans as *Lucifer*. Venus remains in the morning sky for the remainder of 2025, coming to superior conjunction in early January 2026. Its climb above the eastern horizon is rapid for those seeking it from equatorial or southern latitudes; it remains rather low for early risers further north. It is found in the vicinity of a much-dimmer Saturn

Morning Apparition of Mercury
24 March to 30 May

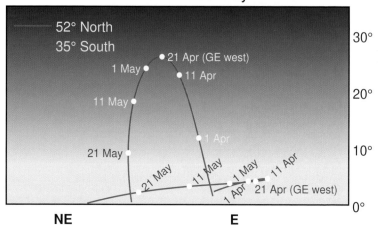

during the first week of April. Retrograde motion ends mid-month, bringing Venus back to Saturn in a much closer pass nearer the end of April.

Earth intercepts a stream of cometary debris late in the month, with the Lyrids peaking around 22–23 April. The Moon will be in its waning crescent phase. See *Meteor Showers in 2025* for more information. The Full Moon of 13 April has the smallest apparent diameter of any Full Moon this year, with apogee taking place 22.5 hours following the instant of Full Moon. (For the largest Full Moon of the year, see the November *Sky Notes*.) Several lunar occultations of stellar targets take place this month: the Pleiades on both 1 April and 29 April, Spica on 13 April, and Antares three days later. Regulus (α Leonis) is found less than 2° away from the waxing gibbous Moon on 8 April.

Mars passes 0.4° south of fourth-magnitude κ Geminorum on the third day of the month. Nine days later, on 12 April, Mars leaves Gemini behind and re-enters the constellation of Cancer. Aphelion occurs on 16 April when the red planet is at a distance of 1.6661 au from the Sun. East quadrature takes place on 21 April with Mars appearing distinctly gibbous through the telescope. Mars is visible in the evening sky, setting in the early morning hours for northern observers but vanishing in late evening for those located south of the equator.

Morning Apparition of Venus
March 2025 to January 2026

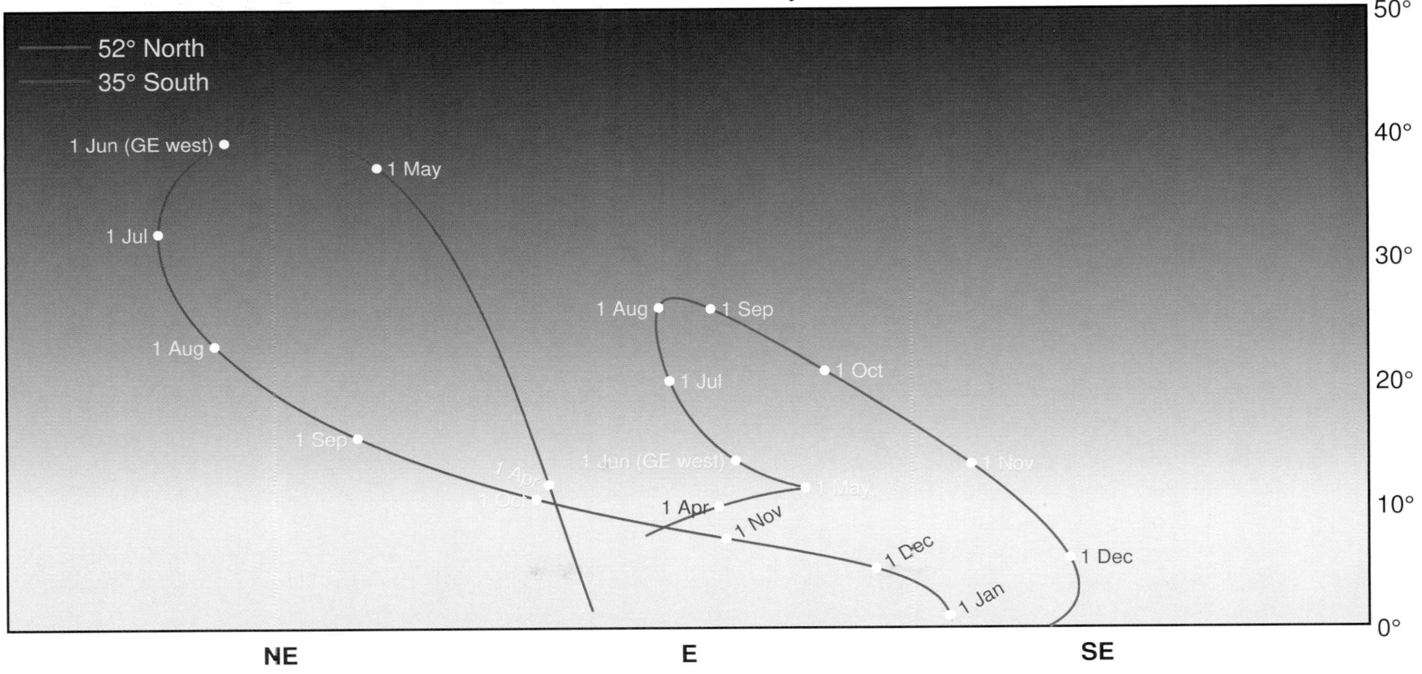

52° North
35° South

1 Jun (GE west)
1 May
1 Jul
1 Aug
1 Sep

1 May
1 Jul
1 Aug
1 Sep
1 Oct
1 Nov
1 Dec
1 Jan

1 Aug
1 Sep
1 Jul
1 Jun (GE west)
1 May
1 Apr
1 Nov
1 Dec
1 Apr

50°
40°
30°
20°
10°
0°

NE
E
SE

Jupiter is found in the evening sky, with the best views from northern temperate latitudes where the magnitude −2.1 planet sets around midnight. Planet watchers in the southern hemisphere lose sight of the gas giant in the twilight of early evening. Jupiter is the brightest starlike object in the constellation of Taurus.

Saturn is a morning object, most easily observed from the southern hemisphere where it rises before dawn brightens the sky. Venus makes a distant (about 8°) pass by the planet in the first week and a closer (about 4°) pass late in the month. The waning crescent Moon joins in the fun on 25 April when it is just 2° away from the ringed planet. Having undergone a geocentric ring-plane crossing last month, Saturn now presents the unlit side of its rings to terrestrial observers; the sunlit side will not return until next month's heliocentric ring-plane crossing. Saturn moves from Aquarius to Pisces on 19 April and will remain in the fishy constellation until retrograde motion returns it to Aquarius in late September.

Uranus is approaching conjunction with the Sun next month and is setting ever earlier in the evening. The sixth-magnitude planet is located in Taurus but is largely lost in evening twilight by the end of the month.

Neptune continues its slow progress through Pisces, now visible before sunrise and best viewed from southern latitudes. Mercury zips past Neptune on 6–7 April, just 0.7° away from the distant planet while the waning crescent Moon is found 1.6° north of the eighth-magnitude body on 25 April. As always, a telescope is necessary to see this eighth-magnitude object.

An Introduction to Unusual Observatory Domes
The (Peculiar) Porter Turret Telescope

Katrin Raynor

My article 'An Introduction to Unusual Observatory Domes: Mills Observatory', published in the *Yearbook of Astronomy 2022*, describes the Mills Observatory in Dundee, Scotland, an unusual astronomical observatory with a 7-metre rotating dome constructed from papier-mâché and one of only two surviving such domes in the UK. In the very traditional and conservative world of astronomy, unusual dome designs are few and far between, although there are two rather wonderful examples located on the eastern side of the United States of America. I will introduce both of these domes over two short articles (see also 'Leo Scanlon's Aluminium Dome' elsewhere in this volume), intertwining the people and events that inspired these wonderful pieces of astronomical engineering. The first of these is the (rather peculiar) Porter Turret Telescope.

Vermont is a state in the New England region of the United States, and is famous for its fabulous autumns and cold winters with temperatures averaging around −6 degrees Celsius over the winter months. For astronomers, preparing to observe during cold clear nights is part and parcel of the hobby, although some may be lucky enough to observe from within the shelter and warmth of an observatory. With these comfort factors in mind, it is said that Arctic explorer and amateur astronomer Russell W. Porter was inspired to construct his famous Porter Turret Telescope, a 12-inch Newtonian reflector that looks and functions quite differently to other more conventional telescopes cocooned inside traditional observatories.

Russell Williams Porter (1871–1949) is often referred to as the founder, or one of the founders, of amateur telescope making. Born in Springfield, Vermont on 13 December 1871, he was well educated, attending Vermont Academy, Norwich University, the University of Vermont, and the Massachusetts Institute of Technology where he gained a degree in architecture. As a young man he travelled to the Arctic eight times before settling back in Springfield in 1919. Already a well-seasoned astronomer and telescope maker, Porter set up a telescope building class in 1920 with the help of his good friend James Hartness, an engineer and politician who was also a telescope builder. Three years later, an astronomy club called the Springfield Telescope Makers was born. They built a clubhouse called Stellafane –

The 10-inch Hartness Turret Telescope, located in the grounds of the nearby Hartness House Inn, Springfield, where it is maintained by the Springfield Telescope Makers. (Richard Koolish)

Latin for 'Shrine to the Stars' – on a 30-acre plot that Porter owned. It remains the location for the annual Stellafane convention to this day.

The Porter Turret Telescope, located on Breezy Hill in Springfield, always draws a lot of attention at the annual convention. The telescope is of course named after Russell W. Porter, whilst the word 'turret' originates from the design which was first fabricated by James Hartness. The telescope has a familiar Newtonian design but does not look or function like a 'normal' Newtonian reflector. For a start, unlike ordinary telescopes the diagonal is situated *before* the primary, and the whole telescope is mounted *outside* of the dome!

The dome itself was constructed using concrete and rotates in right ascension on a steel equatorial ring and rollers. Movement of the dome is applied using a motor which is controlled by a 'speed reduction gear' and friction roller.

A conventional Newtonian telescope reflects light using a parabolic (or curved) primary mirror which collects light from the sky. The gathered light is reflected to a secondary flat mirror which re-directs it to an eyepiece where astronomical objects can be viewed by the observer. The Porter Turret Telescope operates in almost the opposite way in that light hits the flat secondary mirror first before being re-directed through 90 degrees to the primary mirror situated at the end of the struts

The famous and iconic 12-inch f/17 Porter Turret Telescope, pictured here with the distinctive pink Stellafane Clubhouse in the background. (Richard Koolish)

of the telescope. This flat mirror rotates around an axis that connects the eyepiece to the primary mirror which changes the declination. Light is directed through the eyepiece, which sits inside the observatory protruding through a hole in the secondary mirror enabling observers to star gaze whilst keeping warm. The whole telescope moves together in right ascension.

Porter devised and built an array of unusual telescopes, including the beautifully designed Porter Garden Telescope – innovative and ornamental telescopes produced for the garden and which were commercialised by the Springfield-based Jones & Lamson Machine Company in the 1920s. Made from bronze and decorated with ornate floral designs, they could be permanently situated outdoors and were designed so that the optics could be removed and kept elsewhere when not in use.

The Porter Garden Telescope was indeed a superbly designed instrument, and surviving examples are the prized possessions of many well to do stargazers. However, there can be no doubt that Porter's lasting legacy is his visually striking Porter Turret Telescope.

For further information about the history of Stellafane, making and designing telescopes and telescope mirrors, details of previous and upcoming Stellafane conventions, and much more, visit the Stellafane website *Stellafane: Home of the Springfield Telescope Makers* at: **stellafane.org**

My thanks are due to Bart Fried and Richard Koolish for their valuable assistance.

May

2	4 Vesta	Opposition	168°
4	Moon	First Quarter	90°
5	Mars, M44 (Praesepe)	0.6° apart	82°
6	3 Juno	Opposition	161°
6	Saturn	Heliocentric ring plane crossing	48°
6/7	Earth	Eta Aquariid meteor shower (ZHR 30)	
10	Moon, α Virginis (Spica)	0.3° apart: lunar occultation	154°
12	Moon	Full	176°
14	Moon, α Scorpii (Antares)	0.3° apart: lunar occultation	163°
17	Uranus	Conjunction	0°
20	Moon	Last Quarter	90°
25	69230 Hermes	Closest approach to Earth	69°
25	Mercury, Uranus	0.1° apart	6°
26	Moon, M45 (Pleiades)	0.5° apart: lunar occultation	7°
27	Moon	New	5°
30	Mercury	Superior conjunction: morning → evening	1°

Dates are based on UT. Lunar occultations of M45 refer to η Tauri (Alcyone). Peak activity dates for meteor showers are estimates. The last column gives the approximate elongation from the Sun.

Mercury is found in the morning sky, best seen from the southern hemisphere and from equatorial latitudes. It is already losing altitude in the east but remains visible until late into the month for those favoured observers. The planet continues to brighten as it heads toward superior conjunction on 30 May. Mercury passes just 0.1° south of Uranus on 24 May but this event takes place only 6° from the Sun. Mercury passes through its ascending node (with respect to the ecliptic) on 26 May and races through perihelion on the last day of the month.

Venus has returned to direct motion and once again passes by Neptune. This time the two planets are just 2.1° apart in the opening days of May. A telescope is required to spot Neptune but the two worlds are over 40° away from the Sun by this time. Venus passes through its descending node on 8 May, moving from north of the ecliptic to south. As the morning star, Venus is best seen from tropical or southern latitudes but it is slowly gaining altitude above the eastern horizon for those further north searching for the bright planet.

69230 Hermes is at its minimum distance from Earth on 25 May, zipping past at a comfortable 0.5 au away. It shines a feeble magnitude +19 in the constellation of Cancer. 69230 Hermes is an Earth-crossing asteroid, a member of the Apollo group, and crosses Earth's orbit (heading away from the Sun) at the end of the month. It is at its maximum elongation from the Sun, just over 90°, in early July. For more information on 69230 Hermes and its interesting history, see *Minor Planets in 2025*.

Earth collides with the remnants of Halley's Comet this month with the Eta Aquariids peaking during the first week of May. The article *Meteor Showers in 2025* has all of the details. The Moon draws a little closer to Regulus (α Leonis) this month, appearing 1.8° north of the star on 5 May. Occultations of Spica, Antares, and the Pleiades occur on 10 May, 14 May, and 26 May respectively, but the occultation of the Pleiades takes place less than 10° from the Sun and will not be observable.

Mars spends most of the month in the constellation of Cancer before moving into Leo on 26 May. The red planet glides through the northern reaches of the open cluster M44 (better known as the Beehive Cluster or Praesepe) on 5 May. The red planet is visible in the evening skies, setting after midnight for late-night planet watchers in northern temperate latitudes but sinking below the horizon mid-evening for southern observers. A solstice occurs on Mars on 29 May, with summer beginning in the northern hemisphere and winter in the south.

4 Vesta comes to opposition in ecliptic longitude on the second day of the month. Opposition in right ascension takes place three days later. The brightest member of the main asteroid belt is only 1.2 au from Earth and shines at magnitude +5.8, just on the edge of naked-eye visibility in very dark skies. Look for it a little south and west of the fourth-magnitude star 16 Librae.

3 Juno also arrives at opposition this month, first in ecliptic longitude (6 May) and then in right ascension (18 May). It is twice as far away from Earth as 4 Vesta and is only a meagre tenth magnitude in the constellation of Serpens (Caput). More information on both of these asteroids may be found in *Minor Planets in 2025* elsewhere in this volume.

Jupiter continues its traversal of the constellation of Taurus this month, passing 2.0° north of Tianguan (ζ Tauri), a third-magnitude emission line B star, on 18 May. The name Tianguan was assigned to this star by the International Astronomical Union in 2016. It is derived from the name of a Chinese asterism, 天關 or Tiānguān, meaning 'Celestial Gate', which includes this star. Jupiter is very low in the west at sunset for southern hemisphere observers and sets during evening twilight for those further north. Conjunction with the Sun takes place next month.

Saturn reaches an equinox on 6 May with spring beginning in the southern hemisphere of the gaseous planet and autumn getting started in the north. Earthly observers have been viewing the unlit side of the rings since the geocentric ring-planet crossing of 23 March but the heliocentric ring-plane crossing of 6 May will see light return to the ring system. For more details on this year's ring-plane crossings of Saturn, see *Saturn At Its Equinox: The Ring Plane Crossing of 2025* following these *Sky Notes*. The gas giant is traversing the constellation of Pisces this month and is visible before sunrise. The best views of this planet continue to be in the southern hemisphere but observing conditions are improving for planet watchers in the north as Saturn is rising ever closer to midnight.

Uranus is at conjunction with the Sun on 17 May and is lost to view this month. Mercury's very close pass on 24 May takes place too close to the Sun to be observable.

Neptune rises before the Sun in the constellation of Pisces. It remains mired in morning twilight for early risers in northern temperate latitudes so it is best seen from southern hemisphere vantage points where it appears in the east shortly after midnight. Venus and Neptune are just over 2° apart during the first few days of May but a telescope will be necessary to spot the eighth-magnitude blue ice giant. The waning crescent Moon is 1.9° north of Neptune on 22 May.

Saturn at its Equinox
The Ring-Plane Crossing of 2025

David Harper

Saturn's rings have been slowly closing since their maximum opening of 27° at the opposition of 2017. This year, the Earth and Sun both pass through the plane of the rings, events that happen once every 14 or 15 years and which are always eagerly anticipated by Saturn-watchers.

As the year opens, the Earth is 4.3° north of the rings, and the Sun is 1.9° north. The Earth is the first to cross the ring-plane, on 23 March at 19^h UT. Unfortunately, this is only 11 days after Saturn's conjunction with the Sun, and so the planet is invisible in the twilight glare. When the Sun crosses the ring-plane on 6 May, Saturn is 48° west of the Sun, but it remains lost in the early morning sky for observers in Europe and North America. Viewers in the southern hemisphere have a better chance to see Saturn ringless against a dark morning sky, an hour or two before sunrise.

The accompanying diagram shows the latitude of the Earth and Sun relative to Saturn's ring-plane throughout 2025. Saturn's conjunction with the Sun in March is marked by an orange vertical line, whilst opposition in September is indicated by a magenta line.

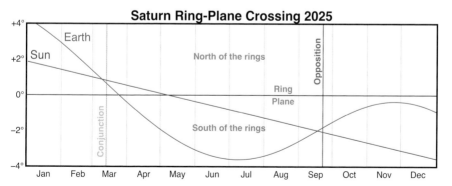

Saturn Ring-Plane Crossing 2025

The latitude of the Sun and Earth relative to the ring-plane of Saturn during 2025. This is the angle at which the rings are tilted towards the Sun or Earth. Saturn's conjunction and opposition dates are marked with vertical lines. (David Harper)

The Sun's latitude relative to the ring plane changes linearly during the year, just as its declination seen from the Earth varies almost linearly during the weeks immediately before and after the terrestrial equinoxes. The Earth's latitude relative to the ring plane follows an S-shaped curve, however. This reflects the Earth's orbital motion. Seen from Saturn, the Earth is an inferior planet, whose motion is retrograde for more than four (Earth) months as it passes through inferior conjunction with the Sun. This is also the period when Saturn's motion is retrograde seen from the Earth, of course.

The diagram reveals that, although the Earth's ring-plane crossing was unobservable, the year offers excellent opportunities to image and sketch the ring system at very low angles. There are parallels with the 2009 ring-plane crossing, which also occurred close to conjunction. Members of the Saturn, Uranus and Neptune Section of the British Astronomical Association (BAA) captured many images of Saturn's vanishing rings during the 2008/9 apparition, which is a "mirror image" of 2025/26.

This year, the rings open quite rapidly between April and June as seen from the Earth, which is 3.6° south of the ring-plane by early July. At this time, the Sun is just 1° below the rings, illuminating them at a very oblique angle. Images taken by BAA members in March and April 2009, when the Earth/Sun/rings geometry

Saturn in 2025

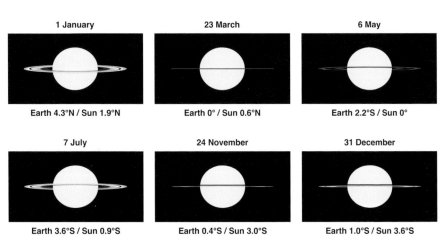

Saturn's appearance on key dates in 2025. 23 March: Earth crosses the rings. 6 May: the Sun crosses the rings. 7 July: greatest opening until 2026. 24 November: minimum opening until the 2038/39 ring-plane crossing. (Lynne Marie Stockman)

was very similar, show that the rings should be easily visible throughout June, July and August.

From 8 July, the rings begin to close up. At opposition on 21 September, the Sun is already 2.0° south of the rings, but the Earth is at 1.8° south, half the angle that it reached in July. The geometry is similar to that of early February 2009, when images taken by BAA members show a distinct black shadow line where the rings crossed the disk of Saturn.

The rings continue to close, reaching a minimum tilt to the Earth of 0.37° on 25 November. This is less than half of the minimum tilt observed in 2008/9, but images taken by BAA members in late December 2008 give a hint of what to expect. In November 2025, as then, the Sun is more than 3° below the ring plane, so whilst the rings may be no more than a knife-edge on either side of Saturn, their shadow on the planet will be visible even in modest telescopes.

The rings remain tilted at less than 1° to the Earth from mid-October until early January 2026, providing observers in all parts of the world with plenty of opportunity to witness this fascinating sight in the evening skies. They then begin to open up as spring returns to Saturn's southern hemisphere. Saturnian mid-summer is in 2032, and we must wait until 2038/39 for the next ring-plane crossing.

June

1	Venus	Greatest elongation west	46°
3	Moon	First Quarter	90°
6	Moon, α Virginis (Spica)	0.5° apart: lunar occultation	128°
9	Earth	Tau Herculid meteor shower (ZHR low)	
10	Moon, α Scorpii (Antares)	0.3° apart: lunar occultation	169°
11	Moon	Full	175°
17	Mars, α Leonis (Regulus)	0.7° apart	64°
18	Moon	Last Quarter	90°
21	Earth	Solstice	
22	Saturn	West quadrature	90°
23	Moon, M45 (Pleiades)	0.6° apart: lunar occultation	32°
23	Neptune	West quadrature	90°
24	Jupiter	Conjunction	0°
25	Moon	New	5°
29	Saturn, Neptune	1.0° apart: conjunction in RA	96°
30	Moon, Mars	0.2° apart: lunar occultation	59°

Dates are based on UT. Lunar occultations of M45 refer to η Tauri (Alcyone). Peak activity dates for meteor showers are estimates. The last column gives the approximate elongation from the Sun.

Mercury once again inhabits the evening sky. This is a fair appearance of the planet for observers in northern temperate latitudes, but it is the best evening apparition of the year for those tracking the planet from the tropics. It is also an excellent apparition for astronomers in the southern hemisphere. Mercury is a bright magnitude −2.1 at the outset of June, gradually dimming to +0.4 by July. It is found 2.0° north (in ecliptic latitude) of Jupiter on 8 June but this will be a difficult planetary conjunction to see with the two worlds just 11° away from the Sun at the time.

Venus reaches its greatest elongation west of 45.9° on the first day of June, with theoretical dichotomy just hours later. For early risers in the tropics and southern hemisphere, Venus reaches its maximum altitude above the eastern horizon this month. However, the bright magnitude −4.2 planet is still ascending for observers in northern temperate latitudes and will continue to climb higher in the morning

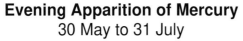

Evening Apparition of Mercury
30 May to 31 July

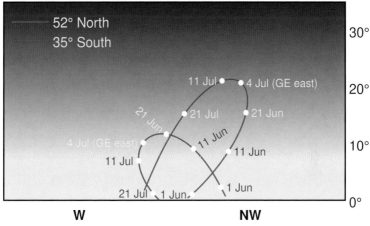

sky for the next two months. In a telescope, Venus appears as waxing gibbous figure that is slowly diminishing in size as it leaves Earth behind in its orbit. Our twin world goes through aphelion on 12 June.

Earth arrives at a solstice on 21 June. On this day the Sun reaches its maximum northerly declination for the year. Summer officially starts in the northern hemisphere and winter claims the south. The waxing gibbous Moon may interfere with observations of the poorly-understood Tau Herculid meteor shower in early June. See *Meteor Showers in 2025* for further details. The waxing crescent Moon is only 1.6° north of Regulus on the second day of June and 1.3° north of the star on 29 June, not yet close enough for an occultation to occur. However, the Moon does occult first-magnitude stars Spica on 6 June and Antares on 10 June. The brightest member of the Pleiades disappears behind the Moon in early morning skies on 23 June. Mars is the final victim, with the red planet vanishing behind our satellite on 30 June.

Mars and the waxing crescent Moon are just over a degree apart on the first day of the month. Later in June, on the seventeenth, Mars passes 0.7° north of Regulus, the brightest star in the constellation of Leo, the Lion. The waxing crescent Moon takes another run at Mars on the last day of the month, this time occulting the rocky world. The end of the lunar occultation event is visible after sunset from the

Triple Conjunction of Saturn and Neptune

Saturn and Neptune come to conjunction three times in right ascension (29 June 2025, 6 August 2025, 16 February 2026) and once in ecliptic longitude (20 February 2026). The two planets were last in conjunction in 1989 and will not meet again until 2061. (David Harper)

western coast of South America. Mars continues to inhabit the evening sky, best seen from northern latitudes where it sets in late evening. Southern hemisphere observers lose sight of the red planet about two hours earlier. Mars continues to diminish in brightness, dimming from +1.3 at the beginning of the month to +1.5 at the end.

Jupiter is at conjunction with the Sun on 24 June and is lost to view for most of the month. Mercury and Jupiter come to within 2° on 8 June but the planets are just over 10° away from the Sun at the time so the event will probably go unobserved. Jupiter moves from Taurus to Gemini on 12 June but its close passes of the stars 1 Geminorum and Propus (η Geminorum) as well as the globular cluster M35 take place too close to conjunction to be visible.

Saturn reaches west quadrature on 22 June which places it 90° away from the Sun. Found in Pisces shining at first magnitude, the ringed planet rises late in the evening for astronomers in the southern hemisphere and just after midnight for those in northern temperate latitudes. On 29 June, Saturn and Neptune undergo their first conjunction in right ascension this year when the two planets appear just a degree apart in the sky. A second conjunction will occur later this year, with a third taking place in 2026.

Uranus emerges from morning twilight by the end of the month for early risers in the southern hemisphere. Look for this faint planet in the constellation of Taurus.

Neptune reaches west quadrature on 23 June and rises before midnight for all observers by the end of the month. Neptune and Saturn are at conjunction in right ascension near the end of June; both planets are found in the constellation of Pisces.

Nightfall
When Darkness Falls

John McCue

In the middle of an open cluster in an undefined galaxy out in the depths of the cosmos there exists Lagash – an exoplanet, as we now call them. In 1941, Isaac Asimov (see Figure 1) created this starting point for his well-known, and very popular, science fiction short story, *Nightfall*. Populated by an intelligent civilisation,

Figure 1: Isaac Asimov, pictured here as a young man without his renowned sideburns. (John McCue).

Lagash revolves around a sun labelled Alpha by the people. The reader will suspect that this name implies that others were nearby, and the reader would be right. Alpha is one of six suns in a multiple system bound together by their mutual gravitational attraction. With so many suns in their sky, the beings of Lagash experience a permanently bright daytime sky on every part of their planet, but the astronomers worked out that a circumstance would happen that enables Darkness to fall once every 2,049 years. Generations of Lagashians pass mythical tales of mysterious things called Stars down the ages, but no-one, with a human-like lifetime, has seen them. The Stars are merely legendary. Asimov's fictional account of the reactions of the Lagash people when they are plunged, en-masse, into a Darkness that no-one alive has experienced, leaves an indelible impression on the mind.

Readers who are familiar with Asimov's *I, Robot* series of futuristic stories will know his powers of prophecy, but Nightfall is surely his masterpiece of prediction. Over 5,000 exoplanets have now been discovered, many of these being members of planetary systems orbiting single stars, like the planets orbiting our sun. But even more intriguing has been the discovery of exoplanets in binary star systems. The planets involved usually orbit just one of the stars, but some orbit both stars; these are known as *circumbinary exoplanets*. Despite this avalanche of exoplanets, their settings pale against Asimov's vision. Currently the record number of stars in which an exoplanet is entangled is four, with two such systems known for certain. The first one was discovered in 2013[1]. This exoplanet nestles in a system known as Kepler-64 AB (*Kepler* being the orbiting survey telescope, and AB being the primary and secondary stars). It was uncovered by the Planet Hunters, citizen scientists who use free public data from the NASA *Kepler* mission. The exoplanet is consequently also known as PH-1b.

Kepler-64 A is a whitish-yellow F-type star, half as big again as our sun in both mass and size. The B star is a red dwarf M-type star, less than half the sun's parameters, this pair lying over 7,000 light years from Earth, and not an easy destination to visit! They circumvent each other in a very tight orbit, much closer than Mercury is to our sun, taking only 20 days to do so. However, a Neptune-sized exoplanet saunters around both of these stars in a wider circuit, at just under Venus' distance from our sun, taking 138 days to do so. It passes in front of star A as it goes, giving away its presence to the Kepler telescope as it looks for dips in a star's light. (Naturally, not all of Kepler's transit dimmings are caused by exoplanets, but many are).

1. See 'Planet Hunters: A Transiting Circumbinary Planet in a Quadruple Star System' by Megan E. Schwamb et al at: **arxiv.org/abs/1210.3612**

There's more … another binary system orbits the Kepler-64 AB star-pair at a huge distance of around 1,000 au, making the whole circus about 25 times larger in scope than our own solar system. No such life, as we know it, is really possible on this planet, but if a civilisation were there, the beings would not even come close to the Lagashian's predicament. AB would always be close in their sky, and the distant pair would be just bright stars. Asimov's prophecy in detail lies in the future, and surely it does, beyond doubt.

A second exoplanet with four parents – 30 Ari[2] – was discovered shortly afterwards, and much closer at 150 light years. The grouping has a similar architecture, except that the planet and a newly-found red dwarf both revolve around the secondary star B, all three objects orbiting the main star A and its companion.

From The Extrasolar Planets Encyclopaedia[3], the author has studied the distribution in distance of known exoplanets that exist in a binary system. He has selected those within 100 parsecs, a distance that would mean there is a good chance that they will be accessible to visual observers, even from built-up areas. The planet will not be apparent in the telescope – that is quite obvious – but to know that one is there, in real time, as you look, stimulates the imagination. From the distance selection, the author has chosen and sketched the exoplanet-harbouring binary system STF 1321, discovered by German astronomer Friedrich Georg Wilhelm von Struve and included in his double star catalogue *Catalogus Novus Stellarum Duplicium et Multiplicum* published in 1827. The exoplanet has a minimum mass of ten times the Earth, making it a super-Earth, and it revolves around B, the secondary star. B is hardly inferior though, being only 0.1 magnitudes fainter than the primary, and orbiting it every 2,000 years at a distance of just over 100 au. The exoplanet is very close to star B, only 0.14 au away, and rounds it every 24 days. Such proximity to the star, coupled with the planet's oversize nature, certainly precludes any possibility of a Lagashian-like population. The primary star A would pose little in the way of light pollution, being so far away!

The author is grateful to Pam Whitfield, who uses a 102mm, f/7 triplet refractor coupled with a monochrome CCD camera which has 5.4 micron pixels. Observing from Silsden in the West Riding of Yorkshire, she imaged STF 1321, giving the opportunity for comparison with the author's sketch. In Pam's image (see Figure 2) the field stars labelled with magnitudes (quoted from the software Sky Chart) were also visually sketched by the author (Figure 3), who used the software AstroImageJ

2. See 'Know the Star, Know the Planet. III. Discovery of Late-Type Companions to Two Exoplanet Host Stars' by Lewis C. Roberts Jr. et al at: **arxiv.org/abs/1503.01211**

3. Visit The Extrasolar Planets Encyclopaedia at: **exoplanet.eu**

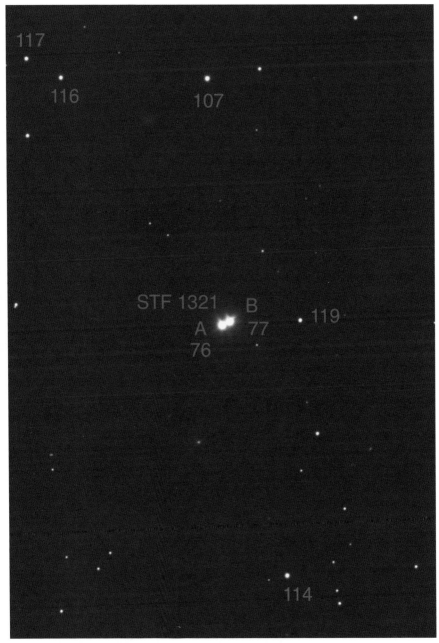

Figure 2: Image of the binary star system STF 1321. (Pam Whitfield)

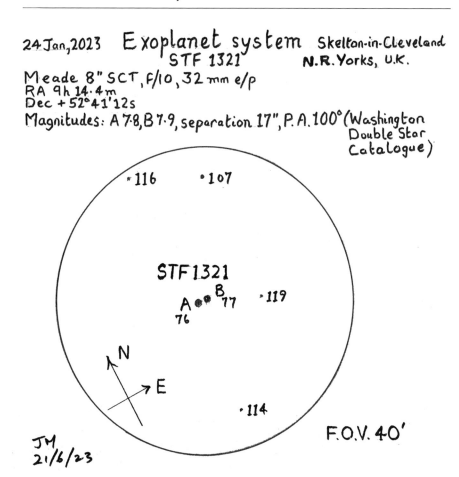

Figure 3: Sketch of the binary star system STF 1321. (John McCue)

to measure, from Pam's image, the separation of the pair at 17.4″ with an error of 0.2″ and the position angle as 99.7° with an error of 0.4°. These values agree well with those given in the *Washington Double Star Catalogue*.

Probably the most comfortable binary-and-exoplanet system to observe from just about anywhere in the world is 83 Leo, whose exoplanet orbits the secondary star, which is easily the fainter of the two. With a swing around the B star in two-and-a-half weeks, the exoplanet is 25 times as massive as the Earth. This mass makes the existence of life unlikely – at least that with which we are familiar – and its close proximity to its star (which is just a little cooler and smaller than our sun) adds to that problem.

July

2	Moon	First Quarter	90°
3	Earth	Aphelion	
3	Moon, α Virginis (Spica)	0.7° apart: lunar occultation	102°
4	Mercury	Greatest elongation east	26°
7	Moon, α Scorpii (Antares)	0.4° apart: lunar occultation	144°
10	Moon	Full	176°
18	Moon	Last Quarter	90°
20	Moon, M45 (Pleiades)	0.8° apart: lunar occultation	58°
24	Moon	New	4°
25	134340 Pluto	Opposition	176°
26	Moon, α Leonis (Regulus)	1.2° apart: lunar occultation	25°
28	Moon, Mars	1.3° apart: lunar occultation	48°
28/29	Earth	Delta Aquariid meteor shower (ZHR 20)	
31	Moon, α Virginis (Spica)	0.9° apart: lunar occultation	76°
31	Mercury	Inferior conjunction: evening → morning	5°

Dates are based on UT. Lunar occultations of M45 refer to η Tauri (Alcyone). Peak activity dates for meteor showers are estimates. The last column gives the approximate elongation from the Sun.

Mercury is at its highest in the west at the beginning of the month, with greatest elongation east (25.9°) occurring on 4 July. Mercury enthusiasts in northern temperate latitudes will lose the tiny planet around mid-month but it remains visible for longer the further south you travel. Aphelion (the third this year) takes place on 14 July, with Mercury entering into retrograde motion a few days later. Inferior conjunction happens very late on the last day of the month.

Venus adorns the morning skies. It continues to gain altitude above the eastern horizon for northern temperate latitude observers but is already descending back toward the ground for astronomers in equatorial and southern regions. In a telescope it appears as a waxing gibbous globe, increasing from 64% illumination to 75% over the course of the month. At the same time, the overall apparent size is decreasing, from 18.0″ down to 14.5″. This is because it is getting more distant from Earth, starting at 0.94 au at the beginning of July and ending 1.16 au away by the end of the month. Venus is 2.4° south of Uranus on 4 July but optical aids will be necessary to spot sixth-magnitude Uranus in dawn skies.

Earth reaches its furthest distance from the Sun on the third day of the month. Just hours later, the Moon occults the first-magnitude star Spica. Occultations of Antares (7 July), the Pleiades (20 July), and Regulus (26 July) follow, with a second occultation of Spica occurring on the last day of the month. For good measure, the Moon also occults Mars one final time this year, on 28 July. The Delta Aquariid meteor shower makes its annual appearance near the end of the month. The Moon is in its waxing crescent phase and should provide no interference. The article *Meteor Showers in 2025* has a description of this event.

Mars continues its slow approach to the Sun, visible in the evening sky until mid- to late evening, depending upon the observer's latitude. On 28 July, Mars crosses from Leo to the constellation of Virgo, and is occulted by the waxing crescent Moon one final time. However, this event is visible only from a small portion of Antarctica. Look for the red planet in the west, shining at magnitude +1.6, as soon as skies turn dark.

Jupiter underwent conjunction late last month and is visible low in the east at dawn. Located in Gemini, the gas giant's close pass by third-magnitude Tejat (μ Geminorum) on the fourth day of the month takes place too close to the Sun to be observable. However, Jupiter rapidly distances itself from the Sun and is rising during the early morning hours by the end of July.

Saturn continues to pull away from the Sun and rises before midnight for all observers this month. The rings have been opening up since the geocentric ring-plane crossing in March and reach a shallow maximum of −3.6° on 7 July. They will then close again until they reach a shallow minimum in November, after which they will begin to widen again. Found in Pisces, first-magnitude Saturn enters into retrograde motion mid-month.

Uranus is a difficult object for planet watchers in northern temperate latitudes as the planet remains mired in morning twilight. Early birds in the southern hemisphere have much the best views of this faint planet which is located in Taurus. For these observers, Uranus is rising just after midnight by the end of the month. On 4 July, look for Uranus and Venus together as they will be just 2.4° apart.

Neptune now rises in the evening, earlier for southern hemisphere observers than for those in the north. A telescope is necessary to locate it in Pisces but it is not far from the very much brighter Saturn. Neptune begins retrograde motion on 5 July.

130340 Pluto reaches opposition in ecliptic longitude on 25 July and in right ascension the following day. Located over 34 au away from Earth, it is only magnitude +15 in the constellation of Capricornus. More information about this demoted planet may be found in *Minor Planets in 2025* elsewhere in this volume.

Mariner 4 Reveals Mars

David M. Harland

Almost sixty years ago, in July 1965, a small robotic spacecraft revealed that most of what we thought we knew about the planet Mars was incorrect.

★   ★   ★

Mars has fascinated astronomers since the invention of the telescope first revealed dark patches and white polar caps on its ochre surface. When nineteenth-century observers perceived narrow linear features crossing the Martian terrain, and named them *canali*, these were widely interpreted as being channels excavated by the natives to transport water from the polar caps to the arid equatorial regions. Although most astronomers rejected these as optical illusions, and Martian engineers as imaginary, it was thought reasonable that the seasonal variations represented the growth cycles of vegetation. It was natural that as soon as it became feasible, we would send a spacecraft to take a closer look.

Mariner 4 was launched by NASA in late-1964 on a trajectory which would fly by the planet on 15 July 1965. It was equipped with a number of science instruments, the primary one being the camera installed on a platform which, as the spacecraft approached the planet, slewed until a light sensor detected its presence. The first picture, taken from a range of 16,500 kilometres, therefore showed the limb against the blackness of space.

Each image frame was an array of 200 by 200 pixels, each encoded as a 6-bit greyscale measurement. These were stored on magnetic tape. In all, 22 pictures were taken, with the final ones beyond the terminator in darkness. After the encounter was complete, the tape was replayed. At 8.33 bits per second, it required 8 hours and 20 minutes to transmit the 240,000 bits in a single frame. The associated engineering information extended the time to 10 hours.

When the raw images were examined, scientists at the Jet Propulsion Laboratory of the California Institute of Technology in Pasadena found that most of them were bland due to 'flare' in the optics. However, they had a computer algorithm which could 'stretch' the contrast of digital images. The cratering which this processing revealed was interpreted as evidence that the surface must be billions of years old, like the Moon, and not at all like the tectonically active Earth.

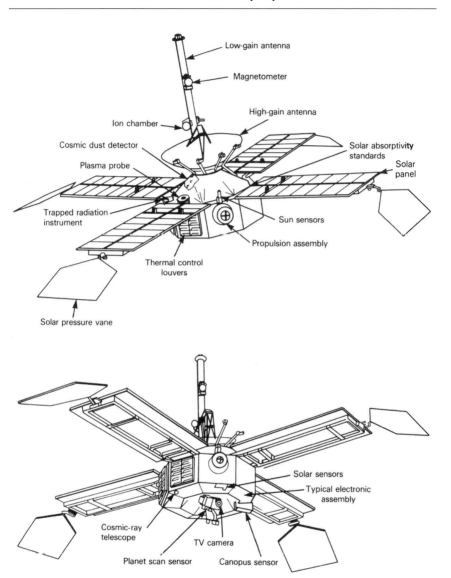

Details of the *Mariner 4* spacecraft. (NASA/JPL-Caltech)

The strength of the spacecraft's radio signal was carefully monitored as it flew behind the planet's limb and again as it emerged from the far side an hour or so later, to profile atmospheric refractivity in daylight at a southern latitude and in darkness at a northern latitude. This allowed the chemical composition, temperature, and

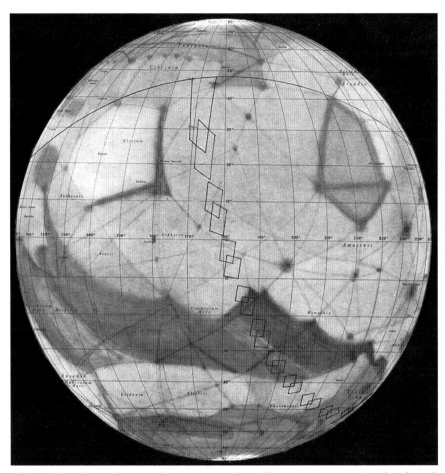

A hemispherical map showing the imaging swath. Unlike astronomers, NASA placed north at the top. The sequence started with an oblique view over the local horizon at about 45°N and ran south and east to about 52°S before crossing the terminator into darkness. Although the 'canali' were widely dismissed as illusory, NASA included a number of them on this presentation. Overall, the imagery covered only about 1% of the planet's surface. (NASA/JPL-Caltech)

pressure of the atmosphere to be inferred. The results indicated surface pressures in the range 4 to 6 millibars. Carbon dioxide in the atmosphere was first identified by the McDonald Observatory in Texas in 1947. A 1964 analysis by the Mount Wilson Observatory in California estimated that carbon dioxide would contribute 4.5 millibars to the surface pressure. Although the spacecraft data confirmed this, astronomers had expected the primary constituent to be nitrogen, which was not so.

Telescopes fitted with thermocouples had measured the temperature of features falling rapidly as they crossed the terminator into darkness, and it had been calculated they might fall to $-100\,^{\circ}C$ by local midnight. But Mars proved to be much colder. Although it had little water vapour in absolute terms, the cold meant that at night the atmosphere was near to its 'saturation point'. It had been expected that the polar caps would be water ice, but the extreme chill revived a suggestion that even if the permanent caps were water ice, the seasonal ones were a frost of carbon dioxide. It turned out that there was a state of equilibrium between the carbon dioxide in the atmosphere and that in the polar caps.

Other instruments on the vehicle established that if Mars possessed a magnetic field, it was too weak to produce a magnetosphere to ward off the solar wind. The irradiation of the surface by energetic particles and solar ultraviolet light severely dented the chances of there being plant life.

In view of the rarefied atmosphere, the seemingly inert ancient crust, and the inhospitable surface, *The New York Times* made a play on Mars' moniker of the 'Red Planet' by naming it the 'Dead Planet'. Overnight, public interest plummeted.

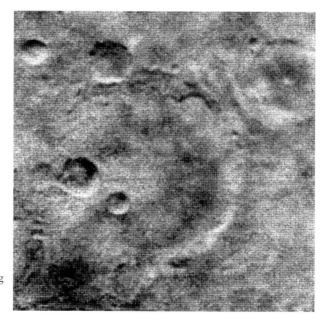

This 120 kilometre wide crater on frame #11 became the iconic image of the mission. It was named Mariner Crater by the International Astronomical Union in homage to the pioneering spacecraft. (NASA/JPL-Caltech)

August

1	Moon	First Quarter	90°
4	Moon, α Scorpii (Antares)	0.5° apart: lunar occultation	117°
6	Saturn, Neptune	1.1° apart: conjunction in RA	133°
7	2 Pallas	Opposition	148°
9	Moon	Full	177°
12	Venus, Jupiter	0.9° apart	36°
12/13	Earth	Perseid meteor shower (ZHR 80)	
13	Earth	Kappa Cygnid meteor shower (ZHR 3)	
16	Moon	Last Quarter	90°
16	Moon, M45 (Pleiades)	0.9° apart: lunar occultation	84°
19	Mercury	Greatest elongation west	19°
23	Moon, α Leonis (Regulus)	1.2° apart: lunar occultation	2°
23	Moon	New	2°
24	Uranus	West quadrature	90°
27	Moon, α Virginis (Spica)	1.0° apart: lunar occultation	50°
31	Moon	First Quarter	90°
31	Moon, α Scorpii (Antares)	0.6° apart: lunar occultation	92°

Dates are based on UT. Lunar occultations of M45 refer to η Tauri (Alcyone). Peak activity dates for meteor showers are estimates. The last column gives the approximate elongation from the Sun.

Mercury returns to the morning sky in what is the best dawn apparition of the year for early birds in northern temperate latitudes. As is usually the case, this is simultaneously the worst morning apparition for observers in the southern hemisphere. Mercury brightens rapidly, from sixth magnitude at the beginning of the month to −1.3 by the end. Retrograde motion ceases on 10–11 August and Mercury's smallest greatest elongation west of the year (18.6°) occurs on 19 August. Mercury passes through its ascending node (relative to the plane of the ecliptic) on 22 August and reaches perihelion five days later.

Venus is just 0.9° south of Jupiter on 12 August but at magnitude −4.0, easily overpowers Jupiter's −2.0. Although the phase of the planet is increasing (75% to 84% over the course of the month), its increasing distance from Earth more than compensates and Venus remains at magnitude −4.0 throughout August. For those

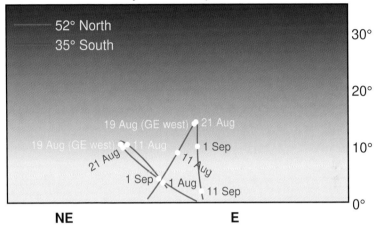

in northern temperate latitudes, the morning star reaches it maximum altitude above the eastern horizon but for those nearer to and south of the equator, Venus has been on the descent for the past month or two. Venus returns north of the ecliptic plane on 29 August when it passes through its ascending node.

Earth is denied a good view of the annual Perseid and episodic Kappa Cygnid meteor showers, with the waning gibbous Moon flooding the sky with light during peak meteoric activity around 12–13 August. For details, see *Meteor Showers in 2025*. The Moon runs through its list of occulted stellar bodies this month, beginning with Antares (4 August), the Pleiades (16 August), Regulus (23 August), Spica (27 August), and Antares one more time at the end of the month. Note that the occultation of Regulus takes place too close to the Sun to observe. The Moon reaches apogee twice this month, with the first one, occurring on 1 August, being the smallest of the year. The Moon also attains the phase of First Quarter twice in August but unlike a second Full Moon in a month, a second First Quarter Moon is not bestowed with a special name.

Mars is an evening sky object, setting early to mid-evening. At magnitude +1.6, it is getting more difficult to spot in the west during twilight. On the second day of the month, Mars is just 0.1° south of Zavijava, the fourth-magnitude beta star in the constellation of Virgo. It passes 0.9° south of another named star, fourth-magnitude Zaniah or η Virginis, on 15 August.

2 Pallas is at opposition in ecliptic longitude on 7 August, having reached opposition in right ascension on 30 July. It lies 2.5 au distant from Earth and is a magnitude +9.5 object in the constellation of Delphinus. More information about 2 Pallas and other members of the asteroid belt may be found in the article *Minor Planets in 2025*.

Jupiter and Venus, the two brightest planets in the sky, pair up on 12 August when they are less than a degree apart in the morning sky. Jupiter is a bright magnitude −2.0 this month but is easily out-shown by the morning star. The largest planet in the solar system rises a little after midnight in the constellation of Gemini. On 18 August, it is found 2.0° north of Mekbuda (ζ Geminorum), a fourth-magnitude Cepheid variable star.

Saturn is in retrograde this month and glides past Neptune for the second time this year, undergoing a second conjunction in right ascension with the ice giant on 6 August. Saturn is magnitude +0.9 but a telescope will be necessary to see eighth-magnitude Neptune just 1.1° away. Saturn is located in Pisces and rises early to mid-evening.

Uranus reaches west quadrature on 24 August and is rising around midnight. The sixth-magnitude planet is located in Taurus and is best seen during the latter part of the month when the Moon is absent from the morning sky.

Neptune rises during the early evening hours but the lengthy twilight of northern temperate latitudes will not facilitate the viewing of this faint planet. It is found near Saturn in the zodiacal constellation of Pisces, and the two gas giants undergo a second conjunction in right ascension early in the month. A third and final conjunction in right ascension (along with a single conjunction in ecliptic longitude) will take place next year. A chart showing the Saturn–Neptune conjunction appears after the June *Sky Notes*.

Gone But Not Forgotten
Anser

Lynne Marie Stockman

The Summer Triangle – comprised of the three first-magnitude stars Altair (α Aquilae), Deneb (α Cygni) and Vega (α Lyrae) – is a well-known asterism in the north celestial hemisphere. It is visible on the meridian around midnight during the northern hemisphere summer and can be seen in the evening hours late into the autumn. The asterism is also visible from much of the southern hemisphere although it is more properly a 'Winter Triangle' for those observers. In the middle of the triangle, south of Albireo (β Cygni) and north of Altair, lie two faint constellations, Vulpecula and Sagitta. The constellation of the arrow, Sagitta, was described by Greek mathematician and astronomer Claudius Ptolemy (c.100–c.170) in his second-century astronomical treatise known as the *Almagest*. The other figure, however, is of much more recent origin.

Although the surrounding constellations are well-defined, the stars in the middle of the Summer Triangle are fourth-magnitude and fainter. For this reason, the region never garnered much attention from the world's ancient astronomers. In traditional Chinese astronomy, this part of the sky resided in the Heavenly Market Enclosure and the first two mansions in the Symbol of the Black Tortoise. For Hindu astrologers, the area was approximately contained in the *nakshatras* (lunar mansions) of Abhijit and Śravaṇa. The three bright stars of the Summer Triangle were well-known to the astronomers of the Middle East but most of the fainter stars within received no special attention. And so it was in western Europe until Polish astronomer Johannes Hevelius (1611–1687) decided to remedy the situation.

Hevelius was a man of many talents. He inherited and ran the family brewery, and was involved in the civic administration of the city of Danzig/Gdańsk, but he is best known for his astronomical accomplishments. He was a skilled observer, discovering the Moon's libration in longitude as well as four comets. He also devised ten new constellations – Canes Venatici, Cerberus, Lacerta sive Stellio, Leo Minor, Lynx sive Tigris, Mons Maenalus, Scutum Sobiescianum, Sextans Uraniae, Triangulum Minus, Vulpecula cum Anser – seven of which still exist today in one form or another.

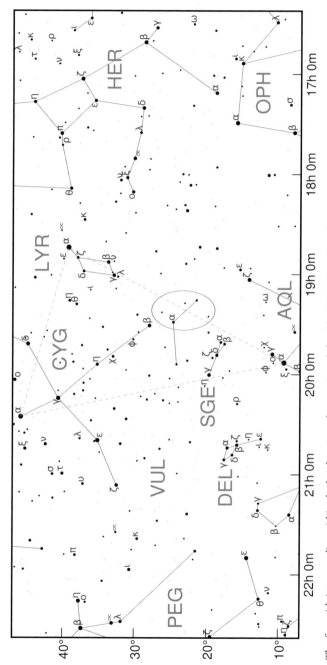

The fox with its goose lies within the bright triangular asterism defined by Altair, Deneb and Vega. Anser (circled in red) is best seen in the evening between August and October. Stars are shown to magnitude +6.0 and the boundaries of the constellation of Vulpecula are shown in cyan. (David Harper)

These new constellations were introduced in the atlas *Firmamentum Sobiescianum sive Uranographia*, contained within the larger work *Prodromus Astronomiae* which was published posthumously by his widow in 1690. On page 117 of *Prodromus Astronomiae* Hevelius explained his reasoning for placing a fox (with a goose) in the region near the existing constellations of Aquila and Lyra. He described the fox as greedy, rapacious and a thief, and that it was in good company with an eagle (Aquila) and a vulture (Lyra was sometimes identified as such). He also said it was carrying the goose to Cerberus, the three-headed dog of Hades and another of Hevelius' creations, which was about to be clubbed senseless by Hercules. The constellation contained 27 stars including a new star which appeared in 1670. This new star was Nova Vulpeculae 1670, also known as CK Vulpeculae or Anthelme's Star, although the exact nature of this cataclysmic variable is still a mystery.

Hevelius named his constellation 'Vulpecula cum Anser', little fox with goose, although the picture in his atlas of a fox with a goose dangling in its jaws was labelled individually as 'Vulpecula' and 'Anser'. Subsequent celestial cartographers labelled the constellation in a variety of ways, sometimes giving it one name in the text and another name on the actual map. Here are a few examples:

Name	Cartographer	Date	Text	Map
Vulpecula cum Anser	J. Hevelius	1690	✔	
	F. W. A. Argelander	1843	✔	
Vulpecula et Anser	J. Hill	1754	✔	
	J. E. Bode	1801	✔	
	A. Jamieson	1822	✔	
	H. M. Bouvier Peterson	1855	✔	
Vulpecula, Anser (separately)	J. Hevelius	1690		✔
	J. Hill	1754	✔	
	J. Flamsteed	1792		✔
	J. E. Bode	1801		✔
	A. Jamieson	1822		✔
Vulpecula	F. W. A. Argelander	1843		✔
	S. Newcomb	1878	✔	
Vulpes	R. A. Proctor	1872	✔	✔
Vulpis	J. Hill	1754	✔	

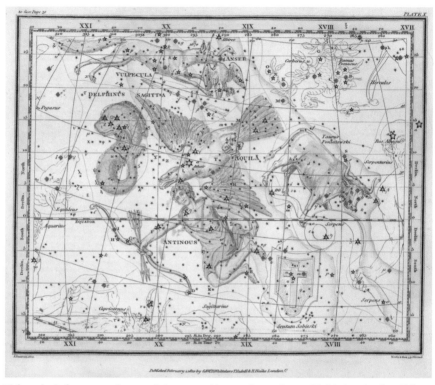

Vulpecula dashes across the top of the page with poor Anser clamped in his jaws in plate X from Alexander Jamieson's 1822 *Celestial Atlas*. (Wikimedia Commons / Alexander Jamieson / United States Naval Observatory Library)

The celestial fox and goose combo appeared widely although not everyone was a fan. William Croswell, who in 1810 produced the first American star map, *A Mercator Map of the Starry Heavens*, ignored most of Hevelius' offerings, introducing instead his own constellations of Scirius Volans, the flying squirrel, and Marmor Sculptile, the bust of Christopher Columbus. Although the fox was always depicted with a goose in various stellar atlases, the name 'Anser' began to fade away, and when the International Astronomical Union (IAU) met in 1922 to organise the sky into official constellations, the fox survived but 'Vulpecula cum Anser' was streamlined to simply 'Vulpecula'.

The little fox, however, still has its prize. In June 2017, the IAU approved the name *Anser* for α Vulpeculae, which, although only magnitude +4.45, is the brightest star in the constellation. Anser is a red giant of spectral type M, with a *Hipparcos*-measured parallax of 10.97 milli-arcseconds which yields a distance of 91 parsecs or nearly 300 light years from the Sun. As long ago as 1971, it was suspected by American astronomer Olin Eggen of sharing a similar motion to a number of other stars, including Arcturus (α Boötis), 29 Ceti, κ Gruis, RT Hydrae and 56 Virginis. The Arcturus moving group is a collection of over four dozen stars which seem to be travelling through space together and are thought to be physically associated. The origin of this group is unclear. Spectroscopy has disproved the idea of a dissolving open cluster but it is possible that this stream of stars is the remnant of a dwarf galaxy which was absorbed into the Milky Way long ago. Other hypotheses involve internal resonances or other perturbations within a bar or spiral arm of our galaxy. Research into the origins and dynamics of moving groups is ongoing.

Anser lies a little over 3° south of Albireo, the colourful double star of Cygnus, and is best viewed in the summer and autumn from the northern hemisphere when Vulpecula leaps high overhead.

Further Reading

Allen, Richard Hinkley, (1963), *Star Names: Their Lore and Meaning*, Dover Publications, Inc.

Balantine, John C., (2016), *The Lost Constellations: A History of Obsolete, Extinct, or Forgotten Star Lore*, Springer Praxis Books.

Ridpath, Ian, (2018), *Star Tales: Revised and Expanded Edition*, The Lutterworth Press.

September

7	Moon	Full: total lunar eclipse	180°
12	Moon, M45 (Pleiades)	1.0° apart: lunar occultation	110°
13	Mercury	Superior conjunction: morning → evening	2°
14	Moon	Last Quarter	90°
19	Moon, α Leonis (Regulus)	1.2° apart: lunar occultation	27°
19	Venus, α Leonis (Regulus)	0.4° apart	27°
19	Moon, Venus	0.7° apart: lunar occultation	27°
21	Saturn	Opposition	177°
21	Moon	New: partial solar eclipse	1°
22	Earth	Equinox	
23	Neptune	Opposition	179°
23	Moon, α Virginis (Spica)	1.0° apart: lunar occultation	23°
27	Moon, α Scorpii (Antares)	0.6° apart: lunar occultation	65°
29	Moon	First Quarter	90°

Dates are based on UT. Lunar occultations of M45 refer to η Tauri (Alcyone). The last column gives the approximate elongation from the Sun.

Mercury soon vanishes from view in morning twilight, undergoing superior conjunction on 13 September. It reappears in the west after sunset, in what is the best evening apparition of 2025 for astronomers in the southern hemisphere. However, the planet is almost undetectable from northern temperate latitudes, not even reaching 10° in altitude. Mercury encounters its descending node on the last day of the month, moving south of the ecliptic.

Venus, the waning crescent Moon, and Regulus form a threesome in the western sky on 19 September. The Moon occults Venus first. This event begins around 10:30 UT and is visible from northern Canada, the Arctic, and northern Russia, near the Kara Sea. Shortly afterwards, the Moon occults Regulus. Venus and Regulus have their closest approach (0.4° distant) at around 16:00 UT. All three objects are 27° from the Sun but Venus is higher above the horizon at sunrise when viewed from the northern hemisphere than from the south.

Evening Apparition of Mercury
13 September to 20 November

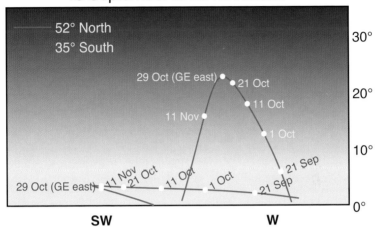

Earth is treated to a total lunar eclipse on 7 September, followed two weeks later by a partial solar eclipse. Observing circumstances are outlined in the article *Eclipses in 2025*. An equinox takes place on 22 September, marking the return of spring in the southern hemisphere and the beginning of autumn in the north. The waning gibbous Moon occults the Pleiades on 12 September. One week later, on 19 September, the waning crescent Moon occults the morning star, with Regulus disappearing behind the Moon's disk shortly thereafter. The occultations of Spica and Antares take place on 23 and 27 September respectively but these events take place in the evening sky instead of at dawn.

Mars is found in the west after sunset, travelling through the sprawling constellation of Virgo. At magnitude +1.6, it is getting more difficult to see at dusk, and sets early in the evening. The red planet passes through its descending node on 23 September; it will remain south of the ecliptic for the remainder of the year.

Jupiter continues its journey across Gemini, passing 0.2° north of Wasat (δ Geminorum) on 6 September and 0.9° north of the planetary nebula C39 (NGC 2392) on 20 September. On 18 September, Jupiter moves through its ascending node and will be found north of the ecliptic until January 2032. Jupiter rises around midnight and is best viewed from the northern hemisphere.

Saturn reaches opposition on 21 September. It is at its closest to Earth on this day and shines a brilliant magnitude +0.6 in the constellation of Pisces. The rings are tilted toward Earth at a narrow −1.8°. The ringed planet rises at sunset and is visible all night. Retrograde motion takes it back into Aquarius on the penultimate day of the month.

Uranus is found in the constellation of Taurus and enters into retrograde motion (relative both to right ascension and ecliptic longitude) on 6 September. It rises in mid- to late evening and is well-place for viewing in both hemispheres.

Neptune reaches opposition on 23 September, rising at sunset and setting at sunrise. It is at this time when it shines at its brightest, magnitude +7.8, and appears as a fuzzy blue disk just 2.53″ across in a telescope. Neptune is found in Pisces not far from Saturn.

An Introduction to Unusual Observatory Domes
Leo Scanlon's Aluminium Dome

Katrin Raynor

The American astronomer, engineer and polar explorer Russell Williams Porter was an inspiration to many. His passion and knack for communicating his love for telescope making and astronomy left a lasting impression on many amateur astronomers, Leo J. Scanlon being a notable example. After reading an article in *Scientific American*, Scanlon was inspired to build his own telescope, along with what would eventually become known as the world's first aluminium dome observatory. The aluminium dome construction would go on to become a model for the manufacture of future observatory domes.

Leo Scanlon. (John McCue)

Born in Pittsburgh, Pennsylvania in 1903, Leo Scanlon started his career as a stenographer before turning his hand to plumbing, a trade in which he was employed for over 55 years. His interest in astronomy, which would be a life-long passion, was sparked when, at the age of seven, he observed Halley's Comet at its return in 1910. He purchased his first telescope in 1920 and a few years later – inspired by Porter – went on to build his first telescope. Scanlon ground his own 6-inch mirror and mounted the scope using plumbing fittings.

With his interest in astronomical observing and telescope making, Scanlon turned his attention to forming a telescope making club. In 1929 his dream became a reality, and the Amateur Telescope Makers of Pittsburgh – a name later changed to the Amateur Astronomers Association of Pittsburgh (AAAP) – was formed. Prospective new members were invited by hand-written invitation from Scanlon and fellow telescope maker Chester Bertram Roe to join the club. The first meeting was held on 9 June 1929 and was attended by four members.

A joyful Leo Scanlon pictured sitting atop the newly-constructed aluminium dome, which he did in order to demonstrate the strength of the dome. One of the objections to building an all-aluminium astronomical observatory dome had been that aluminium was not strong enough to be used for a structure such as this. Leo conclusively proved otherwise. (AAAP Archives)

The idea for an aluminium dome came to Scanlon after he and other AAAP members installed a permanent 10-inch telescope that had been donated to the association and which he had mounted on the roof of his workshop at his home. The mounting became known as the Springfield Mount. As the site was plagued by light pollution, a dome would reduce the effect of the streetlights. There were also potential safety issues, although a dome would provide a safer space for observers. Scanlon and his colleagues decided that this would be an exciting opportunity to construct a new and innovative type of observatory dome using aluminium sheeting. The use of aluminium was an attractive idea due to it being a lightweight material with a good degree of malleability; the cut-rate price of the metal was also a consideration.

The Valley View Observatory took only two and half months to construct. Ten people were involved in the project, their duties ranging from sourcing the aluminium to drawing up the overall design. The dome was to be mounted upon a wooden board and batten frame – thin strips of wood moulding placed over the seams of panel boards – comprising four sides. Flooring was laid, walls constructed, a door installed and a flat roof built. Then followed the difficult jobs of fabricating the track for the rotating dome and making the aluminium dome itself. Two 12-inch circular iron rings were produced which would be fitted for the roller mounting frame at the base of the dome and on the roof of the wooden-framed building. After the completion of the track, the task of assembling the dome was ready to be put into action.

With the assistance of a mathematical professor (name unknown) from the Carnegie Institute of Technology, Pittsburgh (now the Carnegie Mellon University

A number of Springfield Telescope Makers assembled outside Scanlon's completed Valley View Observatory on 23 November 1930. The person standing in the doorway wearing a hat is John M. Pierce. Russell W. Porter is to the right with white shirt and holding his hat. George A. Perry is in front of Porter. The identities of the two other men are unknown. (Springfield Telescope Makers, Inc. Archives)

College of Engineering) high quality patterns were accurately calculated and produced, allowing for the precise cutting, manipulation of the soft metal and coupling of the twelve curved panels for the dome. This number of panels was chosen in order to represent each sign of the zodiac, an idea of the professor's. The cutting process proved to be straightforward thanks to both the accuracy of his calculations and the availability of a moveable wooden form on which the panels could be bent and coupled to guarantee the right shape and dimensions. The production of the dome construction proved to be a huge success. The coupling of the panels was completed using clamps and bolts and finished off with double-turned standing seams, leaving the inside of the dome water-tight and unobstructed.

Every observatory dome needs an opening through which to observe the skies. In this case, a slit was made into the already constructed dome and reinforced

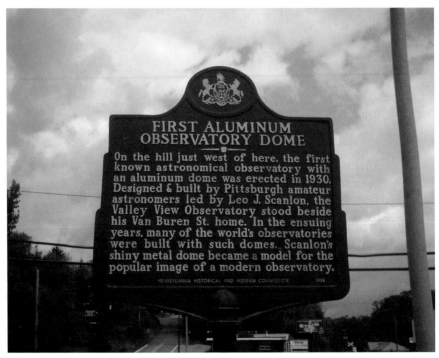

In 1998, the Pennsylvania Historical and Museum Commission erected a marker at the side of a road in McKnight in Ross Township, Allegheny County, Pennsylvania to commemorate Leo Scanlon's aluminium dome. The marker provides a brief description, citing it as "... the first known astronomical observatory with an aluminum dome ..." (Glomar, **HMdb.org**)

with curved pieces of wood. Scanlon fashioned the shutter for the opening using leftover pieces of aluminium, and produced a mechanism to open and close the shutter using a pulley system that worked on brass pipe slide bars.

Weighing in at a hefty 113 kilograms (250 pounds), the dome was ready to be lifted and fitted on top of the wooden building, along with the accompanying task of pairing the dome's rollers on the fixed track. Although a nerve-wracking time, the whole process was completed without any problems, for which Scanlon could not contain his joy.

In order to make the observatory educational and fascinating, the inside of the observatory walls were used to create a star and planet finder. The interior was painted black and approximately 300 stars were painted on to a celestial grid which had panels spaced to represent intervals of 20 degrees in declination and two hours in right ascension. The stars that were chosen for the star finder could be viewed with the naked eye down to third magnitude, and replicated what could be seen in the night sky from the Valley View Observatory site. A small table and red light were installed to enable observers to read star charts.

The Valley View Observatory was dedicated on 23 November 1930 and proved to be an extremely popular destination for keen astronomers over the years. Unfortunately, the observatory was 'torn down' in August 1997, although the aluminium dome was saved with plans to site it atop a new Valley View Observatory. Scanlon's 10-inch telescope now resides at the Hartness-Porter Museum of Amateur Telescope Making, located at Hartness House in Springfield, Vermont.

Further Information

For more information about Leo Scanlon, see the article *The Jovial Genius of Amateur Astronomy: Leo John Scanlon, 1903-1999* by Eric Fischer at:

web.archive.org/web/20110705224337/http://3ap.org/features/leo/leoScanlon Bio1.shtml

An excellent account by Bert Willard of the 1930 (and 1937) Stellafane Conventions, including the Valley View Observatory, can be found at:

stellafane.org/history/early/willard/scanlon.html

October

2	1 Ceres	Opposition	165°
7	Earth	Arid meteor shower (ZHR unknown)	
7	Moon	Full	178°
7/8	Earth	Draconid meteor shower (ZHR 10)	
10	Earth	Southern Taurid meteor shower (ZHR 5)	
10	Moon, M45 (Pleiades)	0.9° apart: lunar occultation	137°
13	Moon	Last Quarter	90°
16	Moon, α Leonis (Regulus)	1.1° apart: lunar occultation	53°
17	Jupiter	West quadrature	90°
21	Moon, α Virginis (Spica)	1.0° apart: lunar occultation	5°
21	Moon	New	3°
21/22	Earth	Orionid meteor shower (ZHR 20)	
25	Moon, α Scorpii (Antares)	0.5° apart: lunar occultation	38°
29	Moon	First Quarter	90°
29	Mercury	Greatest elongation east	24°

Dates are based on UT. Lunar occultations of M45 refer to η Tauri (Alcyone). Peak activity dates for meteor showers are estimates. The last column gives the approximate elongation from the Sun.

Mercury is putting on its best evening show for observers in the southern hemisphere but planet watchers in northern temperate latitudes will need a clear horizon and acute eyesight to spot the tiny world low in the west after sunset. Mercury remains fairly bright throughout the month, from magnitude −0.5 to −0.1, and reaches its greatest elongation east of 23.9° on 29 October. Mercury and Mars are 2° apart in the west on 20 October, and it undergoes its fourth and final aphelion of 2025 earlier in the month on the tenth.

Venus reaches perihelion for the second time this (terrestrial) year on the second day of the month. The morning star is getting lower in the eastern sky every day and is best viewed from northern latitudes. It is at its minimum magnitude of –3.9 this month even as its phase (the illuminated portion of its disk) grows from 91% to 96%. Countering this is Venus's increasing distance from Earth and decreasing apparent size; it is nearly a full arc-second smaller at the end of the month than it was at the beginning.

Earth is bombarded by meteoric material throughout the month, with the Arid, Draconid, and Southern Taurid showers taking place early in October (largely spoiled by the Full or waning gibbous Moon) and the Orionids hitting the atmosphere around the same time as the New Moon. See *Meteor Showers in 2025* for more details about these annual events. Once again the Moon occults four bright stellar objects this month, starting with the Pleiades on 10 October, followed by Regulus on 16 October, and Antares on 25 October. The occultation of Spica on 21 October takes place too close to the Sun to observe but astronomers will get one more chance this year to see α Virginis disappear behind the Moon's disk next month.

Mars is getting to be quite a difficult object to spot in the west as it is deep in evening twilight when it sets. On 1 October, the planet leaves Virgo for the neighbouring constellation of Libra. It is found 0.6° south of Zubenelgenubi (α² Librae) on 14 October but this star is only fifth-magnitude and not visible at dusk. Mercury is just 2° away on 20 October, a sure sign that solar conjunction is approaching. Southern hemisphere observers have the best chances of spotting Mars which is only about 20° away from the Sun by the end of the month.

1 Ceres is the largest of the bodies populating the main asteroid belt between Mars and Jupiter, and was the first minor planet discovered. It arrives at opposition in ecliptic longitude on the second day of the month when it is 2.0 au from Earth and shining at eighth magnitude. Opposition in right ascension occurs on 7 October. Look for 1 Ceres in the constellation of Cetus and read more about this dwarf planet in *Minor Planets in 2025*.

Jupiter reaches west quadrature on 17 October. With shadows cast slightly to one side, Jupiter and its Galilean satellites provide an interesting spectacle through a telescope. Slowly brightening to magnitude −2.3 by the end of the month, the gas giant rises in late evening in the constellation of Gemini.

Saturn is just past opposition and is visible for most of the night, not setting until just before dawn. The first-magnitude planet is located in Aquarius.

Uranus is moving in retrograde across the constellation of Taurus this month. It rises during the early evening hours but at only sixth magnitude, demands dark skies to observe it properly.

Neptune was at opposition late last month so it is already above the horizon as the Sun sets. Found in Pisces, the eighth-magnitude planet is most easily viewed from the northern hemisphere from where it rises high in the sky.

The First 'Hot Jupiter'

David M. Harland

Around thirty years ago, in October 1995, the discovery of the first planet of a solar-type star was announced and confirmed. Astronomers had been making false claims of 'exoplanets' for many years. Ironically, the team that made this discovery wasn't actually seeking planets.

★   ★   ★

Michel Mayor and Didier Queloz of the University of Geneva were interested in objects with masses too large to be planets and too small to be stars. The smallest possible red dwarf is about 8% of a solar mass, or about 80 times the mass of Jupiter. The core of a gas giant that is less massive than this will not initiate nuclear fusion and therefore will not shine in the visible spectrum, but the heat derived from the continuing gravitational collapse will cause it to radiate in the infrared. The lower transition to a planet is at a mass of about 15 times that of Jupiter. In the 1990s astronomers were eager to find these intermediate objects, called 'brown dwarfs'. Since isolated ones would be difficult to detect, Mayor and Queloz decided to try to find them as companions orbiting around normal stars. Armed with a list of 100+ stars, in late 1994 they started to measure their radial velocities for evidence of Doppler shifts caused by unseen companions. This would not be easy, but they were hopeful of a positive result because they had a new instrument.

If a spectrometer is installed on the rear of the telescope it will suffer stress as the telescope changes targets, and even while it is following a single target across the sky. Also, the internal components will adjust to environmental changes. The optical disturbances within the instrument do not normally matter and astronomers are often happy to measure a radial velocity to an accuracy of 1 kilometre per second, but a search for low-mass companions of stars would involve measuring much smaller velocities.

The ELODIE spectrograph was developed by André Baranne at the Marseille Observatory in France. It was not intended to be slung on the end of a telescope, it was to sit in an environmentally controlled room and be fed light by fibre optics. Its diffraction grating was optimised for a high dispersion to enable the Doppler shift

Michel Mayor (left) and Didier Queloz. (University of Geneva)

of a spectral feature to be measured to an accuracy of ±13 metres per second. In 1993 it was installed at the Haute Provence Observatory in France.

Queloz was working for his doctorate with Mayor as his supervisor, and his task was to analyse the data. They started in late 1994 and by March 1995 it was apparent there was a periodicity in 51 Pegasi, one of the brightest stars on their list. The variation of 120 metres per second was unmistakable. What was suspicious was the cycle of only 4.2 days.

Not all radial velocity variations in starlight indicate that the star is wobbling as a result of gravitational perturbations from companions. A star is a balance between its tendency to contract under gravitation and the pressure of radiation escaping from the nuclear reactions that occur in its core. Pulsations might cause the envelope to inflate and contract, imposing a cyclical periodicity on the radial velocity.

Rather than publish immediately, they decided to wait for the star to become visible again in July to verify that the radial velocity cycle was still in phase; it was. They made their announcement at a conference in Florence, Italy, on 6 October and then followed up in November with a paper in *Nature*.[1]

---

1. 'A Jupiter-Mass Companion to a Solar-Type Star', M. Mayor and D. Queloz, *Nature*, **378**, 355–359, 1995.

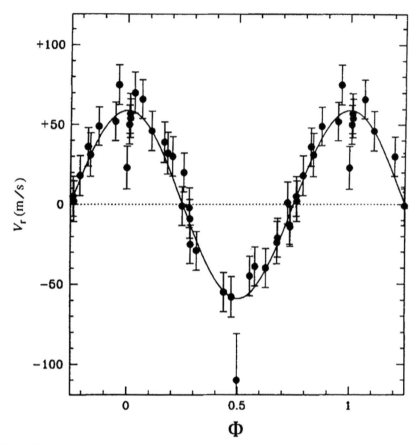

The radial velocity curve in metres per second for 51 Peg. The curve is the best fit to the data, whose error bars are shown. The data has been folded to show the 'phase' (Φ) for an estimated period of 4.2 days. (M. Mayor and D. Queloz, *Nature*, **378**, 355–359, Figure 4, 1995)

Geoffrey W. Marcy and Paul Butler at the University of California at Berkeley had installed a similar spectrograph at the Lick Observatory in California and for some time they had been inspecting nearby solar-type stars for indications of planets, but 51 Pegasi was not one of them. Within a week of hearing of Mayor and Queloz's discovery, they were able to confirm the reported radial velocity signature.[2]

51 Pegasi is about 50 light years from us. It has a spectral type of G5, making it similar to the Sun, which is G2. It is cooler, but its luminosity is greater because its radius is larger. By convention the star is regarded as component 'a' and the first

---

2. *IAU Circular* #6251, 25 October 1995.

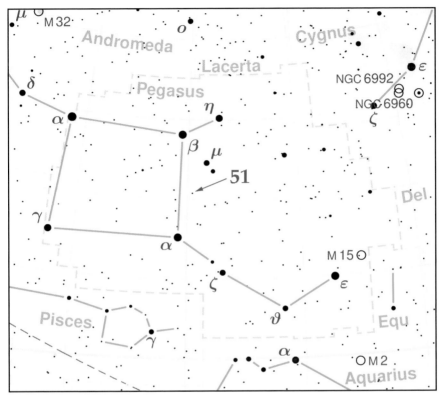

The star 51 Pegasi, formally named *Helvetios* by the International Astronomical Union in 2015, has a visual magnitude of 5.5 and so is a naked-eye object under suitable viewing conditions. (Wikimedia Commons/Torsten Bronger)

companion to be discovered is 'b'. With a mass of roughly half that of Jupiter, it was clearly a planet. Orbiting so close to its parent, at just 0.05 of an Astronomical Unit, its axial rotation is probably tidally locked. The heating of the upper atmosphere on the hemisphere that permanently faces the star will have inflated the radius to a value exceeding that of Jupiter, giving it a very low bulk density. It was the first of a class of exoplanets that were appropriately named 'hot Jupiters'.

Initially referred to as Bellerophon, a mythological figure who rode the winged horse Pegasus, the planet was given the name Dimidium by the International Astronomical Union in 2015 as a Latin reference to its mass being comparable to half that of Jupiter.

In recognition of their discovery, Mayor and Queloz jointly received a half share of the 2019 Nobel Prize for Physics.

November

5	Earth	Southern Taurid meteor shower (ZHR 5+)	
5	Moon	Full: largest apparent diameter (2008″)	176°
6	Moon, M45 (Pleiades)	0.8° apart: lunar occultation	164°
12	Earth	Northern Taurid meteor shower (ZHR 5)	
12	Moon	Last Quarter	90°
12/13	Moon, α Leonis (Regulus)	0.9° apart: lunar occultation	81°
17	Moon, α Virginis (Spica)	1.1° apart: lunar occultation	31°
17/18	Earth	Leonid meteor shower (ZHR varies)	
20	Moon	New	5°
20	Mercury	Inferior conjunction: evening → morning	1°
21	Moon, α Scorpii (Antares)	0.4° apart: lunar occultation	12°
21	Uranus	Opposition	180°
28	Moon	First Quarter	90°

Dates are based on UT. Lunar occultations of M45 refer to η Tauri (Alcyone). Peak activity dates for meteor showers are estimates. The last column gives the approximate elongation from the Sun.

Mercury opens the month at a bright magnitude −0.1 but ends up at sixth magnitude around the time of inferior conjunction on 20 November. Visible in the evening sky at the beginning of the month, it begins retrograde motion on 9 November. This brings it back past Mars three days later but the two planets are only 15° from the Sun at the time. Mercury rises through its ascending node (ecliptic plane) on 18 November, passes between the Earth and Sun on 20 November, and moves through perihelion on 23 November. Mercury joins Venus on 25 November when the two worlds are a degree apart in the morning sky. Retrograde motion ceases on 29 November. This is a decent morning apparition for observers in the northern hemisphere but is relatively poor as seen from southern latitudes.

Venus continues to drop toward the eastern horizon as dawn brightens the sky. Mercury approaches to within a degree of the much brighter morning star on 25 November. Observers in northern and equatorial latitudes have slightly better views of Venus than early risers in the southern hemisphere.

Earth continues to intercept Southern Taurid meteors. A secondary peak occurred last month but the shower is most active around 5 November; unfortunately, the Full Moon will obliterate all but the brightest meteors. The Northern Taurid meteor shower is at its maximum around 12 November although the waning gibbous Moon will likely interfere with observations. Conditions are much better for the famous Leonids which peak around 17–18 November as the Moon will be a waning crescent by then. See *Meteor Showers in 2025* for more details. The Full Moon of 5 November exhibits the largest angular diameter of the year. This coincides with the nearest perigee of the year with the two events occurring just nine hours apart. Interestingly, the largest apogee of the year also takes place this month, on the same day as the New Moon. The waning gibbous Moon occults the Pleiades on 6 November and passes less than 2° north of M44, the Beehive Cluster or Praesepe, on 11 November. An occultation of Regulus takes place over 12–13 November, followed by the final occultation of Spica on 17 November. After this date, the Moon's path will no longer bring it close enough to this first-magnitude star to occult it; this will remain the case until 2031 when a new occultation cycle of Spica begins. The occultation of Antares on 21 November takes place just a day past New Moon.

Mars is very low in the west at sunset and vanishes from view before skies are completely dark. It leaves Libra for Scorpius on the third day of the month and then enters Ophiuchus on 15 November. Mercury is in close attendance on 12 November but the two planets are only 15° away from the Sun. The following day the red planet skims past the globular cluster M80 but it is unlikely that this event will be observable in twilit skies. A Martian equinox takes place on 29 November with spring beginning on the southern hemisphere and autumn starting in the north.

Jupiter rises ever earlier in the evening, now appearing by mid-evening in the constellation of Gemini. The bright planet, shining at magnitude −2.4, enters into retrograde motion on 11 November.

Saturn is well-placed for evening viewing in the constellation of Aquarius, setting a little earlier for those in the southern-hemisphere than for astronomers in the north. The rings have been narrowing since July and reach a shallow minimum tilt of just −0.4° on 24 November. They will widen for the rest of the year. Retrograde motion ceases near the end of the month and the planet resumes its march back toward Pisces.

Uranus reaches opposition on 21 November. It is nearest to Earth at this time (18.5 au) and is at its biggest (3.7″ diameter) and brightest (+5.6 magnitude). Look for it in Taurus any time after sunset; the planet will be visible all night.

Neptune is found in Pisces, appearing in the evening sky and setting around midnight for southern hemisphere observers; it sets somewhat later for those in northern temperate latitudes. It may be found not far from Saturn but a telescope will be necessary to spot the eighth-magnitude planet.

The Moon
The Naming of Full Moons in Native American Culture

Jonathan Powell

As our nearest neighbour in space, the Moon has contributed much to the understanding of the universe. This relates not only to celestial mechanics, but to our own place among the stars. Science aside, the Moon has interwoven itself into the fabric of many cultures across the globe. That colourful tapestry represents a special bond with our lunar companion, a connection that has been handed down from generation to generation.

As the Moon glides effortlessly through its lunar cycle and we observe its ever-changing face, tides ebb and flow, birds migrate, newly born turtles head for the water's edge, and the agricultural world looks skyward for the sign to plant and reap crops. In essence, a watchful Moon maintains a level of synchronicity that, without its very presence and influence, would cause upheaval to much of the life that inhabits the Earth.

The early Native Americans marked time, not by the use of a calendar, but by seasons and the cycle of lunar phases. Whilst some tribes would use numbers, others used names, and despite variance in the wording, an umbilical-like thread would ultimately link an activity or event to a specific time of year and the full Moon that adorned the skies.

However, it is the individual names attributed to the monthly full Moons that capture much of how some cultures perceived and interpreted their life and work on Earth, in conjunction with the Moon itself. Around the world, different peoples connected their own names to each Moon, with the Algonquin tribes, one of the most populous and widespread North American native language groups, gifting their own words for each and every full Moon.

As the year starts, the brilliance of the full Moon and the howl of wolves earned January's full Moon the tag 'Wolf Moon'. February, and the expectation of snowfall, meant that hunting proved difficult and for that reason a 'Hunger Moon' was observed. The Ojibwe people occupying the northern Midwestern United States referring to this Moon as the 'Sucker Fish', an important catch for their winter survival.

The lunar cycle was as important as the passing of the seasons as to how our ancestors went about their daily lives. (Pexels/ Astrofotografias PB)

With spring in the offing, the full Moon in March gained the name 'Worm Moon', as the softening ground saw earthworms strive to break through the layers of soil. Research suggests that the name was only given to this full Moon by Southern Native American tribes, as earthworms did not exist in Northern America at that time. All the earthworms resident in Northern America today are invasive species introduced by colonists. Northern American tribes, such as the Shawnee, therefore called March's full Moon the 'Sap Moon', as maple sap began to flow from the hardwood forest trees.

The influence of the Moon extends beyond tides of the sea, with lakes and rivers also affected by lunar gravitational forces. (NASA/Bill Ingalls)

One of the earliest and most abundant spring flowers, the grass pink or wild ground phlox, earned April's full 'Pink Moon' its name. Other names included 'Sprouting Grass Moon' and, among coastal tribes, 'Fish Moon' when the shad fish swam upstream to spawn. Also, for the new season, eggs were also referenced for the coming of spring. With a whole variety of flowers carpeting the landscape, May's full Moon heralded a solely agricultural naming, 'Full Corn Planting Moon'.

With the ripening of the strawberries ready for picking, June's full Moon became known as 'Rose Moon', with July's 'Buck Moon' appropriately marking the new antlers sported by the buck deer springing forth from their foreheads in coatings of velvety fur. With this time of year being noted for thunderstorms, the tag of 'Thunder Moon' was also applied.

The 'Sturgeon Moon' of August reminded some tribes that during this full Moon, sturgeon in the Great Lakes and other major bodies of water were a substantial food source. A few tribes knew this full Moon as 'Red Moon' because

For several evenings soon after sunset, the bright light of the Harvest Moon illuminates the landscape as the farming community harvest their crops. (NASA/Doug Ellison)

the of the Moon's reddish appearance through the sultry haze of late summer. For the Ojibwe, this Moon marked the season to harvest wild rice.

A time to gather in the crops in September heralded a 'Barley' or 'Harvest' full Moon, and as leaves fell and game became fattened, a 'Hunter's Moon' awaited in October. The Harvest Moon is the full Moon that occurs closest to the autumnal equinox, meaning that it can occur in either September or October.

The need for tribes to keep warm with winter furs saw November's full Moon, the 'Beaver Moon', prompt the setting of beaver traps before the swamps froze over, creating another full Moon name, 'Frost Moon', as temperatures correspondingly dropped.

The year ends with the aptly named full 'Cold Moon' or, among some tribes, the 'Long Nights Moon', both names representing the short hours of daylight, the long winter nights, and perishing cold temperatures.

The generational handing down to subsequent cultures and society of these full Moon names has kept their usage very much alive. The reason for their survival is because in essence, the name still resonates with our perception of the Moon throughout the seasons, with the likes of 'Harvest' and 'Hunter's' full Moon still used in modern dialogue. Perhaps not scientific, but then again, the Moon is not all about science.

December

4	Moon, M45 (Pleiades)	0.8° apart: lunar occultation	167°
4	Moon	Full	175°
7	Mercury	Greatest elongation west	21°
8	Moon, M44 (Praesepe)	1.6° apart	129°
10	Moon, α Leonis (Regulus)	0.7° apart: lunar occultation	108°
11	Moon	Last Quarter	90°
13/14	Earth	Geminid meteor shower (ZHR 75+)	
17	Saturn	East quadrature	90°
18	Moon, α Scorpii (Antares)	0.4° apart: lunar occultation	17°
20	Moon	New	5°
21	Neptune	East quadrature	90°
21	Earth	Solstice	
22/23	Earth	Ursid meteor shower (ZHR 10)	
27	Moon	First Quarter	90°
31	Moon, M45 (Pleiades)	0.9° apart: lunar occultation	140°

Dates are based on UT. Lunar occultations of M45 refer to η Tauri (Alcyone). Peak activity dates for meteor showers are estimates. The last column gives the approximate elongation from the Sun.

Mercury gets brighter over the month and gains altitude above the eastern horizon until around the time of greatest elongation west (20.7°) which occurs on 7 December. From mid-month onward it gets ever lower in the dawn sky. Mercury moves from north to south across the plane of the ecliptic on 27 December.

Venus is below 10° in altitude above the eastern horizon and still descending toward the Sun as it approaches superior conjunction next month. It is nearly fully illuminated but is less than 10″ in apparent diameter. Venus returns to the southern side of ecliptic on 19 December when it slides through its descending node.

Earth is at a solstice on 21 December. The Sun reaches it most southerly declination on this date and begins to head back north. This defines the beginning of summer in the southern hemisphere and the onslaught of winter in the north. Two major meteor showers take place this month, with the waning crescent Moon hopefully

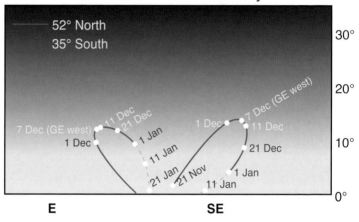

Morning Apparition of Mercury
20 November to 21 January

52° North
35° South

30°

20°

20°

7 Dec (GE west)

1 Dec

11 Dec
21 Dec

1 Jan

1 Dec

7 Dec (GE west)

11 Dec

21 Dec

10°

11 Jan

21 Jan   21 Nov
11 Jan

1 Jan

0°

E   SE

not providing too much interference for the magnificent Geminids around 13–14 December. The Ursid meteor shower later in the month similarly benefits from having a waxing crescent Moon. For more information on these and other meteor showers, see *Meteor Showers in 2025*. The Pleiades are occulted twice this month, on 4 December and again at the end of the month. The waning gibbous Moon banishes Regulus on 10 December but Spica escapes, with the Moon passing 1.3° south of the bright star four days later. The occultation of Antares on 18 December will be difficult to spot as the star is quite close to the Sun by this time.

Mars is less than 10° from the Sun at the beginning of the month and is largely lost to view as it sets shortly after our star. It moves from Ophiuchus to Sagittarius on 10 December, after which is passes close by a number of interesting deep sky objects in the constellation of the Archer, but none of these events will be visible in dusky skies.

Jupiter continues to brighten as it heads toward next month's opposition in the constellation of Gemini. Its retrograde motion takes it back past the planetary nebula C39 (NGC 2392), appearing just over a degree north of the deep sky object on the last day of the month. Jupiter rises in the early evening and is visible until dawn.

Saturn reaches east quadrature on 17 December. This places it firmly in the evening sky, where it sets around midnight for those watching from northern temperate latitudes and an hour or two earlier for astronomers in the southern hemisphere. Look for the first-magnitude planet in the constellation of Aquarius as soon as skies get dark.

Uranus continues it retrograde motion across Taurus this month. On 14 December, it is 0.1° south of 14 Tauri, a G-type giant star of similar brightness to the planet. Eight days later Uranus is 0.2° south of 13 Tauri, a Be-type dwarf star also about the same magnitude as the ice giant. At opposition last month, Uranus is already above the horizon by the time the Sun sets and is visible all evening, setting around midnight for southern hemisphere observers and in the early morning hours for those living in northern temperate latitudes.

Neptune ends its retrogression on 10–11 December when it returns to direct motion across the constellation of Pisces. East quadrature occurs on 21 December. The faint blue ice giant is visible in the evening, setting around midnight for sky watchers in northern temperate latitudes and an hour or two earlier for observers south of the equator. Look for it near the much brighter Saturn.

Rumblings at the Royal Observatory
The Greenwich Power Station Furore

David Harper

In the spring of 1906, the ground beneath the Royal Observatory at Greenwich began to shake, affecting the telescopes that were at the very heart of the Observatory's mission: to provide the world with Greenwich Mean Time. The furore which followed would see letters in *The Times*, a debate in the House of Lords, and a Parliamentary inquiry.

Horse-drawn trams had been a familiar sight in London since the 1860s, but by the early 1900s, the tram network was being converted to run on electricity, and there was a need for a dedicated generating station. Several locations had been considered, but the London County Council decided to use the site of an existing tram depot by the river in Greenwich. This was perfect for the supply of coal by boat. It was an unhappy coincidence that it was only half a mile from the Observatory.

Construction began in 1902, and two tall chimneys soon appeared. At more than 250 feet high, and almost exactly due north of the Airy transit telescope which defines the Greenwich Meridian, the initial concern was that smoke and hot gases rising from the chimneys would interfere directly with observations. It turned out, however, that vibrations from the steam engines would be the real problem.

The huge metal beams of the steam engines were unbalanced, and 94 times every minute, a vertical force of 23 tons alternately lifted and dropped the massive foundation block on which each engine rested, whilst a force of 40 tons moved the block from side to side.

At the Observatory, astronomers began to notice ripples on the surface of the mercury troughs beneath each of the main telescopes. The troughs provided an artificial "horizon" which was used to determine the "level error" of the Airy transit telescope and its neighbour, the altazimuth telescope. This is a measure of how far the altitude axis of a transit telescope deviates from true horizontal, and all transit observations must be adjusted to correct for it.

The Astronomer Royal, William Christie, had noted his concerns about the power station in his annual report to the Observatory's Board of Visitors in May 1906, and this was quickly picked up by many national newspapers. One prominent member of the Board of Visitors, Professor Herbert Turner of Oxford, wrote a

The Royal Observatory at Greenwich was founded in 1675 to provide mariners with accurate data for navigation at sea. Every day since 1833, the time ball has been dropped at exactly 1 p.m. to enable mariners to set their clocks. (Tony Hisgett/Wikimedia Commons)

scathing letter to *The Times*. On 21 June, Lord Ellenborough – a Royal Navy veteran – rose in the House of Lords and delivered an impassioned speech in which he stressed the importance of the Observatory's time-keeping to the seafarers of the British Empire and the world, and – rather unfairly – compared the London County Council to the French anarchist who had tried to blow up the Observatory in 1894. The rest of the debate was more temperate, and included a contribution by the physicist Lord Kelvin, who observed that the vibration problem would be solved if the reciprocating engines were replaced by steam turbines.

A three-man committee was created to investigate the matter on behalf of Parliament, and it sought the advice of Professor William Dalby, FRS, a distinguished engineer. Dalby suggested several methods to reduce the vibrations, including the addition of counterweights, but the committee acknowledged that this solution would be prohibitively expensive. They reported, however, that the Observatory had already found a practical solution to the vibration issue. The saucer-shaped mercury pools had been replaced by flat-bottomed troughs containing a layer of mercury less than half a millimetre deep. This was sufficient to dampen the vibrations from the power station.

The committee also obtained an assurance from the London County Council that when the generating capacity of the power station was increased in the second phase of construction (completed around 1910), turbines would be installed instead of reciprocating steam engines, and these turbines would be used during the night, when the demand for power was light.

The Greenwich Power Station later became part of the electricity supply system for the London Underground, supplementing the Lots Road power station in Chelsea. Since 1998, the London Underground network has been powered directly from the National Grid, and Lots Road was shut down in 2002. The Greenwich power station remains operational as London Underground's emergency backup generator, using modified Rolls-Royce jet engines which can be brought up to full speed in a matter of minutes if the National Grid supply should fail.

The Airy transit telescope was de-commissioned in 1954 after more than a century of observations, and it can still be seen in its original home at Greenwich.

The four chimneys of the Greenwich Power Station are clearly visible from the top of Greenwich Hill, almost exactly due north of the telescopes of the Royal Observatory. (The wub/Wikimedia Commons)

Comets in 2025

Neil Norman

To many astronomers the sight of a comet with a long tail is one of great fulfilment and excitement. The last comet to be easily visible to the unaided eye was C/2020 F3 (NEOWISE) which sported a lovely dust tail visually in the summer of 2020 as the world was locked down during the early stages of the Covid-19 pandemic.

A number of comets have since passed us by, including comet C/2022 E3 (ATLAS) which reached naked eye visibility in early-2023. This comet, though intrinsically fairly average, managed to grab the media headlines with it being a 'green comet last seen by the Neanderthals' some 50,000 years earlier. This would not have been the case due to the comet's low absolute magnitude value of 10.8 although such headlines did at least help to promote comets to the general public.

Discovered by the Zwicky Transient Facility (ZTF) on 2 March 2022, C/2022 E3 (ZTF) is a long-period comet from the Oort cloud. This image was captured by Michael Jäger (Austria) on 4 January 2023 at 5.00 UT, around a week before its perihelion passage on 12 January 2023 carried it to within a distance of 166 million kilometres from the Sun. The green coma comprised of diatonic carbon is clearly visible, as is the short, stumpy white dust tail and the very prominent blue ionic gas tail. (Michael Jäger)

Unless another comet is discovered soon, 2025 looks set to be a year of disappointment as regards the chances of seeing an object brighter than ninth magnitude.

For the very latest comet discoveries it is advised that the reader regularly pays a visit to the British Astronomical Association (BAA) Comet Section website at **www.ast.cam.ac.uk/~jds** which is regularly maintained and updated with all the latest discoveries and news.

Some 74 comets are expected to return to perihelion this year, with no fewer than 63 of these being periodic comets that have passed the Sun on many numerous occasions, meaning they are visually faint due to their volatile ices and gases being expended during previous returns. The remaining objects are broken down to three long-period comets, four comets designated with the D/ prefix (meaning that they are lost), and the remaining four objects being those that are asteroidal in appearance, but on orbital paths similar to comets. However, they will not become anything of general excitement nor great brightness.

With the technical capabilities available to the modern amateur astronomer, I thought it may be interesting to look at the three brightest of the four comets with the D/ prefix ...

18D Perrine-Mrkos

The first of these is 18D Perrine-Mrkos, which was originally discovered by American-Argentine astronomer Charles Dillon Perrine at Lick Observatory, California on 9 December 1896, but considered lost from 1909 until it was rediscovered in 1955 by Czech astronomer Antonín Mrkos. The period is 6.75 years, although the comet can pass close to Mars and Jupiter with the latter potentially having gravitational effects on it. The comet was not observed in 1975, 1982, 1989, 1995, 2002, 2009 or 2017. However, it is always worth looking for, location details for 18D Perrine-Mrkos being as follows.

DATE	RA	DECLINATION	MAGNITUDE	CONSTELLATION
1 Jan 2025	23 46 00	+06 19 15	16.6	Pisces
15 Jan 2025	00 24 15	+07 26 40	16.8	Pisces

D/1886 K1 (Brooks 1)

This comet was discovered on 23 May 1886 by American astronomer William Robert Brooks, who described it as being large, nearly round and weakly diffuse. The comet was last seen in July of that year and an orbit of 5.44 years was determined. However, it has never been seen since.

DATE	RA	DECLINATION	MAGNITUDE	CONSTELLATION
1 Apr 2025	15 23 03	−16 26 06	?	Libra
15 Apr 2025	15 21 59	−18 22 05	?	Libra
1 May 2015	15 13 27	−20 37 29	?	Libra

D/1884 O1 (Barnard 1)

This comet was discovered by Edward E. Barnard on 16 July 1884 as it was travelling through the constellation of Lupus and situated just 0.42 au from Earth. Although the discovery magnitude is not known, the comet was probably at around 9.5 indicating a very weak object given the short distance from our planet. Peaking at magnitude 8 in August of 1884, the comet unexpectedly flared up in September of that year, although it was never to be seen again. The following table gives the best-known location details for this comet.

DATE	RA	DECLINATION	MAGNITUDE	CONSTELLATION
15 Oct 2025	16 52 58	−27 27 49	?	Scorpius
1 Nov 2025	18 00 07	−26 44 01	?	Sagittarius
15 Nov 2025	18 56 57	−24 40 59	?	Sagittarius

Best Prospects for 2025

At the time of writing, just three comets are expected to attain magnitude 11 or slightly brighter during 2025, details of which are as follows.

21P/Giacobini-Zinner

Discovered by French astronomer Michel Giacobini on 20 December 1900 and recovered by German astronomer Ernst Zinner in 23 October 1913, 21P/Giacobini-Zinner is a member of the Jupiter family of comets and has an orbital period of 6.6 years. It is also the parent body of the Draconid – previously known as the Giacobinid – meteor shower, which is active between 6 and 10 October each year. During the apparition of 2018 spectra revealed that the comet is depleted in carbon/chain molecules and carbon dioxide, strongly suggesting it formed in a warmer part of the solar system.

DATE	RA	DECLINATION	MAGNITUDE	CONSTELLATION
1 Mar 2025	22 38 52	+01 50 09	14	Aquarius
15 Mar 2025	23 38 55	+03 30 55	14	Pisces
1 Apr 2025	00 52 56	+05 15 23	11	Pisces

Comet 21P/Giacobini-Zinner as imaged by Michael Jäger (Austria) on 1 July 2018 at 21.07 UT. (Michael Jäger)

33P/Daniel

This object was discovered by American astronomer Zaccheus Daniel on 7 December 1909. A member of the Jupiter family of comets, 33P/Daniel has an orbital period of eight years although it was not recovered during the predicted returns for 1916, 1923 and 1930. Japanese astronomer Shin-ichi Shimizu recovered the comet on the 1937 return, with all returns being observed since then, with the exceptions of 1957 and 1971. Between 11 and 30 January 2009 the comet went into outburst, increasing its brightness by around three magnitudes. It is worth noting that the comet is frequently subject to perturbations from Jupiter, which have had the effect of slightly increasing the orbital period.

DATE	RA	DECLINATION	MAGNITUDE	CONSTELLATION
15 Nov 2024	07 26 52	+33 54 08	11	Gemini
1 Dec 2024	07 31 35	+38 02 16	12	Auriga
15 Dec 2024	07 27 57	+41 52 12	13	Auriga
1 Jan 2025	07 14 55	+45 57 37	12.5	Lynx
15 Jan 2025	07 01 16	+48 18 53	12.5	Lynx

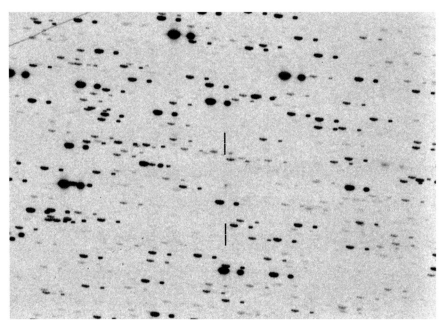

Comet 33P Daniel as imaged by Michael Jäger (Austria) on 6 October 2016. The comet is visible as a faint fuzzy object between the two vertical bars. (Michael Jäger)

24P/Schaumasse

Discovered by French astronomer Alexandre Schaumasse on 1 December 1911, 24P/Schaumasse has an orbital period of 8.24 years, and is also a member of the Jupiter family of comets. The comet was not seen in 1935, 1968, 1976 and 2009, while during the return of 2017 it reached magnitude 10. On 22 March 2010 it passed within 0.02 au of the dwarf planet Ceres, and during the 2025/2026 return it will approach to within 0.6 au (90 million kilometres) of Earth.

DATE	RA	DECLINATION	MAGNITUDE	CONSTELLATION
1 Dec 2025	10 20 57	+19 04 46	12	Leo
15 Dec 2025	11 30 39	+17 16 53	11	Leo
1 Jan 2026	12 51 32	+14 06 35	11.5	Coma Berenices
15 Jan 2026	13 47 57	+11 20 30	11.5	Boötes

Minor Planets in 2025

Neil Norman

Minor planets – often referred to as asteroids – are a collection of varying sized pieces of rock left over from the formation of the Solar system around 4.6 billion years ago. Millions of them exist, and to date almost 800,000 have been seen and documented, with around 550,000 having received permanent designations after being observed on two or more occasions and their orbits being known with a high degree of certainty. Different family types of asteroids also exist, such as Amor asteroids which are defined by having orbital periods of over one year and orbital paths that do not cross that of the Earth. Apollo asteroids have their perihelion distances within that of the Earth and thus can approach us to within a close distance and Trojan asteroids have their home at Lagrange points both 60 degrees ahead and behind the planet Jupiter respectively. These asteroids pose no problems to the Earth.

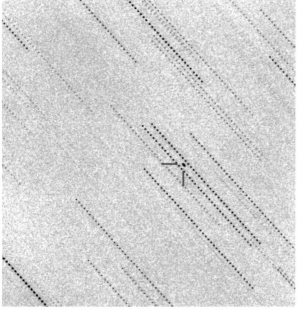

With an estimated diameter of between three and eight metres, the Apollo-type asteroid 2023 BU is ranked as the fourth closest non-impacting asteroid. This image – taken during the close approach of 2023 BU – was captured on 26 January 2023, just hours before the object passed within 3,600 kilometres of the Earth's surface. (Northolt Branch Observatories)

Most asteroids travel around the Sun in the main asteroid belt, which is located between the orbits of Mars and Jupiter. However, some asteroids have orbits which allow them to interact with major planets, including Earth. Over 20,000 asteroids have orbits that can bring them into close proximity to our planet. Such an object is referred to as a Near Earth Asteroid (NEA). To be classed as a Near Earth Asteroid, an object must have an orbit which allows it to pass within a distance of 1.3 au of the Earth. In addition, there are around 2,300 objects – referred to as Potentially Hazardous Asteroids (PHAs) – with orbital paths that can bring them to within an extremely close distance to our planet. A Potentially Hazardous Asteroid is defined as being one which can approach to within 0.05 au (19.5 lunar distances) of the Earth, and have a diameter of at least 140 metres.

Objects of this size could pose a serious threat if on a collision course with Earth. It is estimated that several thousand exist with diameters of over 100 metres, with around 150 of these being over a kilometre across. A large number of smaller asteroids, measuring anything between just a few meters in diameter to several tens of meters wide, pass close to our planet on a regular basis, with considerable numbers of smaller ones entering the Earth's atmosphere every day, burning up harmlessly as meteors.

Those observers with a particular interest in following these objects should go to the home page of the Minor Planet Center. It is their job to keep track of these objects and determine orbits for them. This page can be accessed by going to **www.minorplanetcenter.net** where you will find a table of newly discovered minor planets and Near Earth Objects (NEOs). At the top of the page is a search box that you can use to locate information on any object that you are interested in, and from this you can obtain ephemerides of the chosen subject. The Minor Planet Center site is the one that all dedicated asteroid observers should consult on a regular basis.

Observers who are new to asteroid hunting should remember that these objects will appear as little more than points of light in your field of view. You therefore need to compare your observations with detailed star charts (especially for the dimmer objects) to ascertain that you have indeed observed your intended target. Details of a selection of minor planets observable during 2025 are given below. Some of the objects described are visible in binoculars or small telescopes, and – arriving at opposition during this year – are particularly good targets for the backyard astronomer. Others are much fainter, and will therefore present greater challenges to the would-be observer.

All positional data contained in the following tables is sourced from the JPL Solar System Dynamics website which can be found at **ssd.jpl.nasa.gov**

1 Ceres

This is no doubt the star of the show and arguably the best known of the minor planets. With a diameter of 945 kilometres Ceres is the largest object in the asteroid belt. Discovered by Italian astronomer Giuseppe Piazzi from Palermo Observatory, Sicily on 1 January 1801, Ceres was initially believed to be a planet until the 1850s when it was reclassified as an asteroid following the discoveries of many other objects in similar orbits.

Ceres comes to opposition on 2 October in the constellation of Cetus, reaching a maximum of 165° from the Sun while at a distance from 2 au from Earth.

DATE	RA	DECLINATION	MAGNITUDE	CONSTELLATION
1 Sep 2025	01 19 50	−07 20 29	8.0	Cetus
15 Sep 2025	01 12 40	−08 36 45	7.7	Cetus
1 Oct 2025	01 00 31	−09 58 43	7.5	Cetus
15 Oct 2025	01 48 40	−10 47 04	7.6	Cetus
1 Nov 2025	00 36 18	−10 59 13	7.9	Cetus
15 Nov 2025	00 29 58	−10 27 49	8.2	Cetus

2 Pallas

Pallas has a diameter of 512 kilometres and was the second asteroid to be discovered when first spotted by the German astronomer Heinrich Wilhelm Matthias Olbers on 28 March 1802. He named the object after the Greek goddess of wisdom and warfare Pallas Athena, an alternative name for the goddess Athena. Pallas travels around the Sun once every 1,686 days, its orbit being highly eccentric, and its path around the Sun steeply inclined to the main plane of the asteroid belt, rendering it fairly inaccessible to spacecraft.

Pallas comes to opposition on 7 August in the constellation of Delphinus when at an angular distance of 148° from the Sun and a distance from of 2.5 au from our planet.

DATE	RA	DECLINATION	MAGNITUDE	CONSTELLATION
1 Aug 2025	20 40 06	+15 38 11	9.4	Delphinus
15 Aug 2025	20 28 40	+13 25 27	9.4	Delphinus
1 Sep 2025	20 19 26	+10 35 39	9.5	Delphinus
15 Sep 2025	20 13 59	+07 30 54	9.7	Aquila

3 Juno

With a mean diameter of around 270 kilometres, Juno ranks as the twelfth largest asteroid and is one of the two largest stony (S-type) asteroids, comprising around 1% of the mass of the entire asteroid belt and being the second largest stony asteroid after 15 Eunomia. Discovered by the German astronomer Karl Ludwig Harding on 1 September 1804, its mean distance from the Sun is 2.6 au, its journey around our star taking 4.36 years to complete.

Juno comes to opposition on 6 May in the constellation of Serpens, attaining a distance of 161° from the Sun while a distance from Earth of 2.4 au.

DATE	RA	DECLINATION	MAGNITUDE	CONSTELLATION
1 May 2025	15 54 18	−03 44 40	10.0	Libra
15 May 2025	15 42 26	−02 32 30	9.9	Serpens
1 Jun 2025	15 30 23	−01 45 48	10.0	Serpens
15 Jun 2025	15 20 25	−01 31 08	10.3	Serpens
1 Jul 2025	15 14 04	−01 48 03	10.5	Serpens

4 Vesta

Discovered by German astronomer Heinrich Wilhelm Matthias Olbers on 29 March 1807, Vesta is one of the largest of the asteroids with a diameter of 525 kilometres and an orbital period of 3.63 years. Vesta holds the distinction of being the brightest minor planet visible from Earth. With a maximum magnitude of 6, Vesta is the only one of the minor planets which is regularly bright enough to be seen with the naked eye.

Vesta comes to opposition on 2 May in the constellation of Libra, around 1.2 au from Earth and at a distance of 168° from the Sun.

This image of 4 Vesta was taken on 4 February 2021, one month before its opposition on 4 March of that year. ('Northolt Branch Observatories)

DATE	RA	DECLINATION	MAGNITUDE	CONSTELLATION
1 Jan 2025	13 58 12	−04 53 39	7.8	Virgo
15 Jan 2025	14 19 01	−06 09 11	7.6	Virgo
1 Feb 2025	14 41 36	−07 11 44	7.4	Virgo
15 Feb 2025	14 57 00	−07 36 47	7.1	Libra
1 Mar 2025	15 08 30	−07 37 01	6.9	Libra
15 Mar 2025	15 14 58	−07 13 10	6.6	Libra
1 Apr 2025	15 14 46	−06 17 33	6.2	Libra
15 Apr 2025	15 07 47	−05 20 26	5.9	Libra
1 May 2025	14 54 09	−04 22 53	5.6	Libra
15 May 2025	14 40 49	−03 58 55	5.7	Virgo
1 Jun 2025	14 28 11	−04 19 27	6.0	Virgo

134340 Pluto

Discovered on 18 February 1930 by American astronomer Clyde Tombaugh, and orbiting the Sun once every 248 years at a mean distance of around 5.9 billion kilometres, Pluto was considered as being the ninth fully-fledged planetary member of the solar system until 2006, when it was downgraded to the status of dwarf planet by the International Astronomical Union.

This year Pluto comes to opposition on 25 July in the constellation of Capricornus, with a solar elongation of 176° and at a distance of around 34 au from Earth. For those equipped with large aperture telescopes, and perhaps a great deal of patience, Pluto may be located by using the information given here.

DATE	RA	DECLINATION	MAGNITUDE	CONSTELLATION
1 Jun 2025	20 27 31	−22 52 04	15.0	Capricornus
15 Jun 2025	20 26 37	−22 57 35	15.0	Capricornus
1 Jul 2025	20 25 27	−23 03 45	15.0	Capricornus
15 Jul 2025	20 24 05	−23 10 10	15.0	Capricornus
1 Aug 2025	20 22 37	−23 16 23	15.0	Capricornus
15 Aug 2025	20 21 11	−23 21 59	14.4	Capricornus
1 Sep 2025	20 19 46	−23 27 08	14.4	Capricornus

69230 Hermes

Hermes was discovered on 28 October 1937 by German astronomer Karl Wilhelm Reinmuth on images taken at the Heidelberg-Königstuhl State Observatory.

However, after a brief observational period of only four days the asteroid became too faint to observe and was subsequently lost with no definitive orbit having being determined. Although the object was classed as a lost asteroid at the time, and so did not receive an official number, Reinmuth nevertheless named it after Hermes, the Greek messenger of the gods. It was on 15 October 2003 that American astronomer Brian A. Skiff – who was involved with the Lowell Observatory Near-Earth-Object Search (LONEOS) – observed an asteroid that, once its orbit had been calculated back, proved to be

The recovery image of 1937 UB (Hermes) obtained by Brian A. Skiff using the LONEOS 59-cm Schmidt telescope on 15 October 2003. (NASA / Brian A. Skiff / Lowell Observatory)

the long-lost Hermes. During the 2003 return the object was subjected to radar imaging, and it became clear that Hermes was in fact a binary system with primary and secondary components of 810 metres and 540 metres respectively.

Hermes is an Apollo (Earth-crossing) asteroid, orbiting the Sun once every 778 days, with aphelion and perihelion distances of 2.69 au and 0.62 au respectively, and an orbital inclination of 6° to the ecliptic. Once the orbit had been accurately calculated, it was discovered that Hermes had passed to within just 739,000 kilometres of Earth on 30 October 1937, with an even closer approach of 634,520 kilometres having taken place on 26 April 1942.

A large telescope with CCD equipment will be needed to glimpse Hermes this time around, but for those with the observational capability details of the apparition are given below.

DATE	RA	DECLINATION	MAGNITUDE	CONSTELLATION
1 Apr 2025	02 47 42	+19 33 28	19.0	Aries
15 Apr 2025	04 06 56	+24 59 27	19.0	Taurus
1 May 2025	05 44 53	+27 11 44	19.0	Taurus
15 May 2025	07 45 37	+23 18 45	18.8	Gemini
1 Jun 2025	09 48 24	+12 24 10	19.0	Leo
15 Jun 2025	11 23 48	+00 32 45	19.0	Leo
1 Jul 2025	12 35 04	−08 25 58	19.6	Virgo
15 Jul 2025	13 21 40	−13 38 48	20.0	Virgo

Meteor Showers in 2025

Neil Norman

A shooting star dashing across the sky is a wonderful sight that often captures the imagination of many young minds and perhaps sparks an interest for astronomy in them. On any given night of the year you can expect to see several of these, and they belong to two groups – sporadic and shower. Quite often the ones you see will be 'sporadic' meteors, that is to say they can appear from any direction and at any time during the observing session. These meteors arise when a meteoroid – perhaps a particle from an asteroid or a piece of cometary debris orbiting the Sun – enters the Earth's atmosphere and burns up harmlessly high above our heads, leaving behind the streak of light we often refer to as a "shooting star". The meteoroids in question are usually nothing more than pieces of space debris that the Earth encounters as it travels along its orbit, and range in size from a few millimetres to a couple of centimetres in size. Meteoroids that are large enough to at least partially survive the passage through the atmosphere, and reach the Earth's surface without disintegrating, are known as meteorites.

At certain times of the year the Earth encounters more organised streams of debris that produce meteors over a regular time span and which seem to emerge from the same point in the sky. These are known as meteor showers. These streams of debris follow the orbital paths of comets, and are the scattered remnants of comets that have made repeated passes through the inner solar system. The ascending and descending nodes of their orbits lie at or near the plane of the Earth's orbit around the Sun, the result of which is that at certain times of the year the Earth encounters and passes through a number of these swarms of particles.

The term 'shower' must not be taken too literally. Generally speaking, even the strongest annual showers will only produce one or two meteors a minute at best, this depending on what time of the evening or morning that you are observing. One must also take into account the lunar phase at the time, which may significantly influence the number meteors that you see. For example, a full moon will probably wash out all but the brightest meteors.

The following is a table of the principle meteor showers of 2025 and includes the name of the shower; the period over which the shower is active; the Zenith Hourly Rate (ZHR); the parent object from which the meteors originate; the date of peak

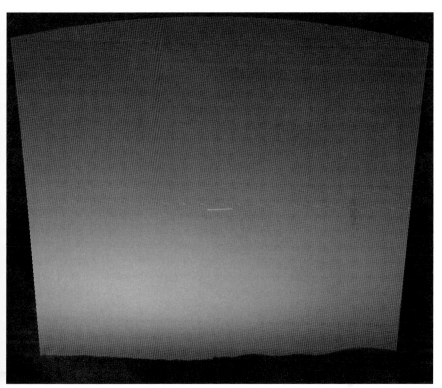

This image, taken on 7 March 2004 by the Mars Exploration Rover-A (MER-A) *Spirit*, shows what is thought to be the first meteor photographed from Mars. (NASA/JPL/Cornell)

shower activity; and the constellation in which the radiant of the shower is located. Most of the information given is self-explanatory, but the Zenith Hourly Rate may need some elaborating.

The Zenith Hourly Rate is the number of meteors you may expect to see if the radiant (the point in the sky from where the meteors appear to emerge) is at the zenith (or overhead point) and if observing conditions were perfect and included dark, clear and moonless skies with no form of light pollution whatsoever. However, the ZHR should not be taken as gospel, and you should not expect to actually observe the quantities stated, although 'outbursts' can occur with significant activity being seen.

The observer can make notes on the various colours of the meteors seen. This will give you an indication of their composition; for example, red is nitrogen/oxygen, yellow is iron, orange is sodium, purple is calcium and turquoise is magnesium. Also, to avoid confusion with sporadic meteors which are not related to the shower, trace the path back of the meteor and if it aligns with the radiant you can be sure you have seen a genuine member of the particular shower.

Meteor Showers in 2025

SHOWER	DATE	ZHR	PARENT	PEAK	CONSTELLAT
Quadrantids	1 Jan to 5 Jan	120	2003 EH$_1$ (asteroid)	3/4 Jan	Boötes
Lyrids	16 Apr to 25 Apr	18	C/1861 G1 Thatcher	22/23 Apr	Lyra
Eta Aquariids	19 Apr to 28 May	30	1P/Halley	6/7 May	Aquarius
Tau Herculids	19 May to 19 Jun	Low	73P/Schwassmann–Wachmann	9 Jun	Hercules
Delta Aquariids	12 Jul to 23 Aug	20	96P/Machholz	28/29 Jul	Aquarius
Perseids	17 Jul to 24 Aug	80	109P/Swift–Tuttle	12/13 Aug	Perseus
Kappa Cygnids	Jun to Sep	3	Unknown	13 Aug	Cygnus
Arids	28 Sep to 14 Oct	?	15P/Finlay	7 Oct	Ara
Draconids	6 Oct to 10 Oct	10	21P/Giacobini–Zinner	7/8 Oct	Draco
Southern Taurids	10 Sep to 20 Nov	5	2P/Encke	10 Oct	Taurus
Orionids	? Oct to 7 Nov	20	1P/Halley	21/22 Oct	Orion
Southern Taurids	10 Sep to 20 Nov	5+	2P/Encke	5 Nov	Taurus
Northern Taurids	20 Oct to 10 Dec	5	2004 TG$_{10}$ (asteroid)	12 Nov	Taurus
Leonids	6 Nov to 30 Nov	Varies	55P/Tempel–Tuttle	17/18 Nov	Leo
Geminids	4 Dec to 17 Dec	75+	3200 Phaethon (asteroid)	13/14 Dec	Gemini
Ursids	17 Dec to 26 Dec	10	8P/Tuttle	22/23 Dec	Ursa Minor

Quadrantids

The parent object of the Quadrantids has been identified as the near-Earth object of the Amor group of asteroids 2003 EH$_1$ which is likely to be an extinct comet. With peak rates known to exceed 100 meteors per hour, the Quadrantids rivals the August Perseids, although there is a drawback in that the period of maximum activity takes place over a very short period of between two and three hours. The radiant lies a little to the east of the star Alkaid (η Ursae Majoris) and the meteors are fast moving, reaching speeds of 40 km/s. Maximum activity occurs on the night of 3/4 January when a waxing crescent Moon setting early in the evening should leave dark skies and the potential for a good show.

Lyrids

Produced by particles emanating from the long-period comet C/1861 G1 Thatcher – which last came to perihelion on 3 June 1861 – these are very fast moving meteors that approach speeds of up to 50km/s. The peak falls on the night of 22/23 April with the radiant lying near the prominent star Vega in the constellation of Lyra. This year, a thin and waning crescent moon should not cause too many problems for observation of this shower, the best viewing being had from a dark location after midnight.

Eta Aquariids

One of the two showers associated with 1P/Halley, the Eta Aquariids are active for a full month between 19 April and 28 May. The radiant lies just to the east of the star Sadalmelik (α Aquarii), from where up to 30 meteors per hour are normally expected during the period of peak activity, although displays of up to 60 meteors per hour can occasionally be seen. Maximum activity occurs in the pre-dawn skies of 7 May when a waxing gibbous moon will tend to block out some of the fainter meteors this time around.

Tau Herculids

Appearing to originate from the star Tau (τ) Herculis, this shower runs from 19 May to 19 June with peak activity taking place on 9 June. The Tau Herculids were first recorded in May 1930 by observers at the Kwasan Observatory in Kyoto, Japan. The parent body has been identified as the periodic comet 73P/Schwassmann–Wachmann, discovered on 2 May 1930 by the German astronomers Arnold Schwassmann and Arno Arthur Wachmann during a photographic search for minor planets being carried out from Hamburg Observatory in Germany.

73P/Schwassmann–Wachmann has an orbital period of 5.36 years. During 1995 the comet began to fragment, and by the time of its 2006 return, at least eight individual fragments were observed (although the Hubble Space Telescope spotted dozens more). 73P/Schwassmann–Wachmann appears to be close to total disintegration. The observed rate of meteors from this shower is low, and with peak activity in 2025 taking place just a couple of days before Full Moon, the chances are that few meteors from this shower will be seen.

Delta Aquariids

Probably linked to the short-period sungrazing comet 96P/Machholz, the Delta Aquariids is a fairly average shower which coincides with the more prominent Perseids. However, Delta Aquariid meteors are generally much dimmer than those associated with the Perseids, making their identification somewhat easier. The radiant lies to the south of the Square of Pegasus, close to the star Skat (δ Aquarii). Located to the north of the bright star Fomalhaut in Pisces Austrinus, the radiant is particularly well placed for those observers situated in the southern hemisphere. The shower peaks during the early hours of 29 July, when a crescent Moon will be setting early in the evening, producing dark skies and the potential of a reasonable display.

Perseids

Associated with the parent comet 109P/Swift-Tuttle – and radiating from a point in northern Perseus, close to the border with the adjoining constellation Cassiopeia – the Perseids are a beautiful sight, with fast moving meteors appearing as soon as night falls and up to 80 meteors per hour often recorded. This is usually one of the best meteor showers to observe, with large numbers of very bright meteors often seen. At the time of peak activity, on the night/morning of 12/13 August, a waning gibbous Moon may obscure all but the brightest meteors this year.

Kappa Cygnids

The Kappa Cygnids is a weak meteor shower, the radiant of which starts at the antihelion point in late June and then locates to Cygnus in July before taking up a position just west of the bright star Vega in early-August. The ZHR is rather poor at just 3 meteors per hour. The shower is named after the position of the radiant at the peak of shower activity, which is near the star Kappa (κ) Cygni. One point of interest is the shower is usually devoid of activity, bursting into what little activity it has once every seven years, with the last recorded activity being that of 2021. In 2025, on the night of supposed maximum activity, the presence of a waning gibbous Moon will ensure poor observing conditions.

Southern Taurids/Northern Taurids

Running collectively from 10 September to 10 December, the Taurids are two separate showers, with southern and northern components, the Southern Taurids linked to the periodic comet 2P/Encke and the Northern Taurids to the asteroid 2004 TG$_{10}$. Meteors from both components emanate from the western regions of Taurus, the radiant for the Southern Taurids being located a little to the north of the star Xi (ξ) Tauri – close to the border with neighbouring Cetus – and that for the Northern Taurids immediately to the south east of the Pleiades open star cluster. The southern hemisphere encounters the first part of the stream, followed later by the northern hemisphere encountering the second part. The ZHR of these showers is low (between 5 and 10 per hour), although they can be beautiful to watch as they glide across the sky. The Southern Taurids appear to peak twice; on 10 October

This image of Comet 2P/Encke was captured on 23 June 2004 by the Multiband Imaging Photometer for Spitzer (MIPS) instrument on the Spitzer Space Telescope and reveals that the path of the comet through space is littered with dust from numerous previous perihelion passages which is seen here spreading away from the comet. (NASA/JPL-Caltech/M. Kelley (Univ. of Minnesota))

when a waning gibbous Moon may leave only the brightest meteors visible, and again on or around 5 November, with rates of between 5 and 10 meteors per hour expected, but with the light of a Full Moon hampering observations. For the Northern Taurid peak on 12 November, the Moon will be at Last Quarter, which means that both showers will be somewhat disappointing this year.

Arids

In 1999 astronomers began speculating that a new meteor shower could be forming, with a radiant in the southern constellation Ara (the Altar) and resulting due to the Earth encountering the debris left behind by the periodic comet 15P/Finlay. In September 2021, the Central Bureau for Astronomical Telegrams (CBAT) released a circular announcing the discovery of the new meteor shower as predicted.

Peter Jenniskens (SETI Institute and NASA Ames Research Center); Timothy Cooper (Astronomical Society of Southern Africa); and Dante Lauretta (University of Arizona), reported that Cameras for Allsky Meteor Surveillance (CAMS) video-based meteoroid orbit survey networks in New Zealand and Chile detected the new meteor shower. Fireballs resulting from the Earth running into the stream left behind from the 1995 return of the comet were first observed on the evening of 28/29 September 2021. Activity lasted into October with our planet encountering the debris left behind by the comet during its 2014/15 return. A total of thirteen slow moving meteors were detected during the period of visibility.

The Arids is a new shower and activity will be unpredictable. Some years will see a limited number of meteors and other years perhaps showing greater activity. This is certainly a meteor shower for observers in the southern hemisphere to monitor closely. As far as the 2025 shower is concerned, at the time of predicted peak activity on 7 October, the Moon will be at full phase, resulting in a poor year for observations.

Draconids

Also known as the Giacobinids, the Draconid meteor shower emanates from debris left behind by the periodic comet 21P/Giacobini-Zinner. The duration of the shower is just 4 days from 6 October to 10 October, with the shower peaking on 7/8 October. The ZHR of this shower varies with poor displays in 1915 and 1926 but stronger displays in 1933, 1946, 1998, 2005 and 2012. Radiating from the "head" of Draco, the meteors from this shower travel at a relatively modest 20 km/s. The Draconids are generally quite faint which, when coupled with the fact that the Moon will be at full phase at the time of peak activity, will result in poor observing conditions with the light from most of the Draconids being swamped in the lunar glare.

Orionids

The second of the meteor showers associated with 1P/Halley, the Orionids radiate from a point a little to the north of the star Betelgeuse in Orion. Best viewed in the early hours when the constellation is well placed, the shower takes place between

2 October and 7 November with a peak on the night of 21 / 22 October. The velocity of the meteors entering the atmosphere is a speedy 67 km / s. This year will present an excellent opportunity for viewing the Orionids, with the New Moon being absent, and leaving dark skies for what may be an excellent show.

Leonids

Running from 6 November to 30 November, this is a fast moving shower with particles varying greatly in size and which can create lovely bright meteors that occasionally attain magnitude −1.5 (or about as bright as Sirius) or better. The

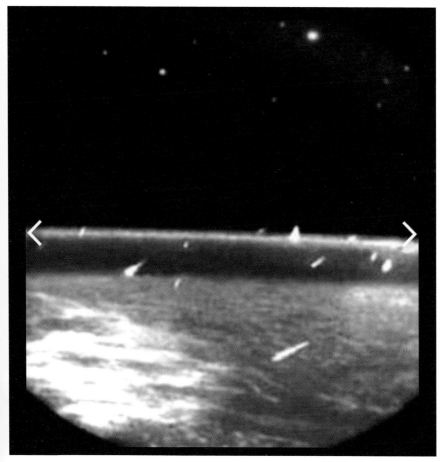

This composite of images taken by the Midcourse Space Experiment (MSX) satellite shows the Leonid meteor shower during its peak in 1997. (Wikimedia Commons / NASA / Leonid Daily News)

radiant is located a few degrees to the north of the bright star Regulus in Leo. The parent of the Leonid shower is the periodic comet 55P/Tempel-Tuttle which orbits the Sun every 33 years. It was last at perihelion in 1998 and is due to return in May 2031. The Zenith Hourly Rates vary due to the Earth encountering material from different perihelion passages of the parent comet. For example, the storm of 1833 was due to the 1800 passage, the 1733 passage was responsible for the 1866 storm and the 1966 storm resulted from the 1899 passage (for additional information see Courtney Seligman's article 'Cometary Comedy and Chaos' in the *Yearbook of Astronomy 2020*). The Leonid shower peaks on the night of 17/18 November when the light from a thin and waning crescent Moon will leave dark skies and the potential for a reasonably good show.

Geminids

The Geminid meteor shower was originally recorded in 1862 and originates from the debris of the asteroid 3200 Phaethon. Discovered in October 1983, this rocky five kilometre wide Apollo asteroid has an unusual orbit that carries it closer to the Sun than any other named object of its type. Classified as a potentially hazardous asteroid (PHA), 3200 Phaethon made a relative close approach to Earth on 10 December 2017, when it came to within 0.069 au (10.3 million kilometres/6.4 million miles) of our planet. The Geminid radiant lies near the bright star Castor in Gemini and the shower peaks on the night of 13/14 December. This is considered by many to be the best shower of the year, and it is interesting to note that the number of observed meteors appears to be increasing annually. This year a Last Quarter Moon will tend to block some of the fainter meteors, although a high ZHR for the Geminids of 75+ should still ensure a good show.

Ursids

Discovered by William Frederick Denning during the early twentieth century, this shower is associated with comet 8P/Tuttle and has a radiant located near Beta (β) Ursae Minoris (Kochab). With relatively low speeds of around 33 km/s, the Ursids are seen to move gracefully across the sky. Research jointly carried out by Dutch/American astronomer Peter Jenniskens and his colleague, the Finnish astronomer Esko Lyytinen (1942–2020), revealed that outbursts may occur when 8P/Tuttle is at aphelion due to some meteoroids being trapped in a 7/6 orbital resonance with Jupiter. The Ursids run from 17 December to 26 December with peak activity taking place on the night of 22/23 December. This year, a thin and waxing crescent moon will have set early in the evening, leaving behind it the dark skies essential for a good display.

Article Section

Recent Advances in Astronomy

Rod Hine

James Webb Space Telescope – First Year of Science

There can be little doubt that the James Webb Space Telescope (JWST), launched on Christmas Day, 2021, has fulfilled and even exceeded all expectations. This image shows the amazing improvement in resolution in the infra-red range compared to the Spitzer space telescope – no mean performer itself.

In July 2023, the NASA's JWST team celebrated the first full year of science with a public on-line event and the release of spectacular images. One of the most notable of these images is of Herbig-Haro object HH 46/47, a region of active star formation about 1,470 light years distant in the southern constellation of Vela.

SPITZER IRAC 8.0μ   WEBB MIRI 7.7μ

NASA/JPL-Caltech (left), NASA/ESA/CSA/STScI (right)

A composite image of HH 46/47 captured by the James Webb Space Telescope showing jets from a pair of young stars, embedded in a disc of gas and dust. (NASA/ESA/CSA/J DePasquale (STScl))

The first Herbig-Haro object was observed in 1890 by the eminent American astronomer Sherburne Wesley Burnham (1838–1921), using the newly built 36-inch Great Lick Refractor telescope. Burnham was actually searching for a different object, already known as Hind's Nebula, but with assistance from Edward Emerson Barnard his discovery was identified as an object in its own right. Originally named Burnham's Nebula, it was not initially recognised as a special class of object, despite its unusual spectrum. During the 1940s, George Herbig (1920–2013) and Guillermo Haro Barraza (1913–1988) independently studied similar objects and realised that they were a by-product of the star formation process. Like Burnham's Nebula, they exhibit strong emission lines of hydrogen, sulphur and oxygen.

In subsequent years, many similar objects were discovered, but in 1982, Michael A. Dopita and I. Evans, from the Mount Stromlo and Siding Springs Observatories, Canberra, Australia, and R. D. Schwartz of the University of Missouri, St Louis, showed that the Herbig-Haro object HH 46/47 was a violent bipolar ejection of jets of matter from a very young star. Their observations were based on photographs from the 4-metre telescope at the Cerro Tololo Inter-American Observatory and much of HH 46/47 is obscured by dark clouds of dust. Their spectra observations in visible light came from the 1-metre telescope at Siding Spring Observatory.

The new high-definition infra-red images from JWST reveal amazing details of the outflows of material, largely hidden from the view of the authors of the 1982 paper. They also show that the jets arise from a pair of interacting stars.

From the Birth of Stars to the birth of a Planet – V960 Monoceros

In 2014, astronomers observed that the star V960 Monoceros exhibited an extreme flare event. Located around 5,000 light years from Earth, its brightness increased by around twenty times and naturally, the flare attracted the attention of many

A composite image of SPHERE data from the VLT and ALMA data. The VLT image in yellow shows the dusty spiral arms, while the ALMA data in blue shows the dense clumps that will form giant planets. (ESO/ALMA (ESO/NAOJ/NRAO)/Weber et al)

astronomers. A group in Chile used the Very Large Telescope (VLT) – a facility operated by the European Southern Observatory – to monitor the star after its outburst. The VLT's Spectro-Polarimetric High-contrast Exoplanet REsearch (SPHERE) instrument revealed that the star was surrounded by an intricate series of spiral arms, rather than the disc-like regions of dust surrounding most young stars.

The spiral arms extend over distances greater than our entire solar system, and work published in 2023 by Philipp Weber of the University of Santiago, Chile, and his colleagues has revealed surprising results. Weber and his team searched the archives of the Atacama Large Millimetre Array (ALMA) for observations of V960 at longer wavelengths and found that within the spiral arms surrounding the star, the dusty matter is collapsing into giant Jupiter-sized lumps. Such a process is likely to happen suddenly as the matter in the spiral arms suffers gravitational instability and is at odds with the commonly accepted view that planets form by core accretion, the gradual build up from smaller objects colliding and joining. This scenario normally occurs over tens of millions of years.

It seems likely that the flare outbursts and the sudden collapse of accreting matter are linked, in which case the clumps could collapse over time scales as short as 20 years. If confirmed, it gives a whole new perspective to the ways that planets are formed.

Missing Dark Matter Puzzle

Not everything pans out smoothly in the world of astronomy. The current best model of our universe calls for specific amounts of dark matter and dark energy, neither of which has yet been positively observed or fully explained. The case for dark matter largely arose as a result of research carried out by American astronomer Vera Rubin (1928–2016) during the 1970s, her work on the spectra of galaxies establishing without doubt that the rotation of galaxies could not be explained without the addition of extra mass to the mass attributable to visible matter.

The adage "the exception that proves the rule" seems appropriate when considering NGC1277, a massive elliptical galaxy located at a distance of around 240 million light years in the Perseus cluster of galaxies, and first discovered in 1875 by Lawrence Parsons, the 4th Earl of Rosse. A galaxy of this type and size – around 150 billion solar masses – would be expected to include anywhere between 10 and 70 % dark matter. Recent work, led by Sebastién Comerón from the Instituto de Astrofísica de Canarias and the University of La Laguna, suggests that NGC1277 cannot have more than 5% dark matter and may even have none. This is completely unexpected, especially as it is a very old galaxy that has not changed much during the last 12 billion years or so. Comerón and his colleagues

The lenticular galaxy NGC 1277 in an image taken by the Hubble Space Telescope. NGC 1277 plays host to one of the most massive black holes known. (NASA/ESA/ Andrew C. Fabian/ Remco C. E. van den Bosch (MPIA))

used a spectrograph with the 2.7-metre Harlan J Smith telescope at the McDonald Observatory, University of Texas, USA. The spectra measurements allow the velocity of the stars within the galaxy to be plotted as a kinematic map, from which they were able to construct a detailed map of the distribution of mass within the galaxy to a radius of more than 20,000 light years. The results clearly indicate a significant lack of dark matter. Furthermore, it must have lost its dark matter very early or even been deficient in dark matter from its formation. No satisfactory explanation has been accepted as to how the dark matter could have been removed or escaped, and alternative theories involving modifications to the laws of gravity are highly controversial.

Hubble Space Telescope Still Making Discoveries

The Hubble Space Telescope (HST) is now well into its fourth decade and still producing beautiful images and fascinating science. This recently released picture of galaxy Z 229-15, some 330 million light years distant in the constellation of Lyra, is just one more challenge to astronomers.

At first sight Z 229-15 looks like a fairly ordinary spiral galaxy, with a bright core and two arms of stars, although further research revealed that it has an active galactic nucleus, or AGN, due to a supermassive black hole in the heart of the galaxy. That causes the central core to be abnormally bright and radiate massive amounts of electromagnetic radiation across a very wide spectrum. The energy comes from the intense heating of the matter in the disc as it is drawn into the black hole by the immense gravitational pull of the black hole.

Situated at a distance of around 390 million light years in the constellation Lyra, Z 229-15 is imaged here in beautiful detail by the NASA/ESA Hubble Space Telescope. (ESA/Hubble & NASA, A. Barth, R. Mushotzky)

The unusual brightness qualifies Z 229-15 to be classified as a quasar.[1] Such objects are very common and because of their brightness can be seen over vast inter-galactic distances, although this one is relatively close to the Earth.

Z 229-15 also qualifies to be classed as a Seyfert galaxy,[2] displaying characteristic emission and absorption lines in the ultra-violet part of the spectrum, very often

1. Extremely luminous active galactic nuclei, also known as quasi-stellar objects.
2. Seyfert galaxies are a type of galaxy named in honour of the American astronomer Carl Keenan Seyfert (1911–1960), who first described them during the 1940s.

with significant variability over time scales from a few hours to a few years. It is now reckoned that around 10% of all galaxies can be classified as Seyfert galaxies. In most other regular galaxies the spectra show predominantly absorption lines due to stars.

NANOGrav

The North American Nanohertz Observatory for Gravitational Waves (NANOGrav) is a collaborative project to use the world's largest radio telescopes to detect the invisible ripples in space-time caused by ultra-massive black holes over cosmic time. The existence of gravitational waves was predicted from Albert Einstein's 1915 theory of General Relativity. Such ripples or gravitational waves were first directly detected in 2014 by the Laser Interferometer Gravitational-Wave Observatory (LIGO), a pair of extremely sensitive instruments located in the USA. The original LIGO observations were of transient ripples caused by the collision of black holes, and the observations were transformed into audio as brief "chirp" sounds.

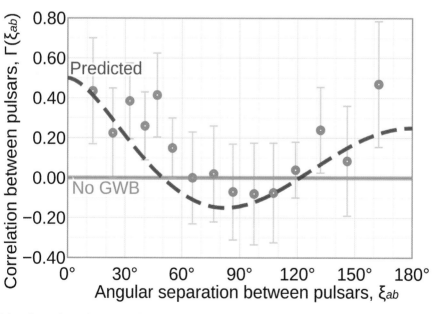

Plot of correlation between pulsars observed by NANOGrav against angular separation between pulsars, compared with a theoretical model (dashed purple). The sold green line shows the prediction if there were no gravitational wave background. (Wikimedia Commons / Cmglee / Attribution-ShareAlike 4.0 International License / CC BY-SA 4.0)

Green Bank Telescope, West Virginia USA. (Wikimedia Commons / NRAO / AUI)

However, the universe is probably awash with small and long-period gravitational waves from the turbulent events at the "Big Bang" and subsequent collisions and mergers of huge black holes. LIGO and other interferometers are unable to detect such long-period ripples, so a different technique is needed. The principle behind NANOGrav is the use of distant pulsars as highly accurate clocks. Pulsars, discovered in 1967 by astrophysicist Susan Jocelyn Bell[3] at Cambridge, are rapidly spinning neutron stars that emit beams of radiation, rather like cosmic lighthouses. The flashes of radio waves, and even light in a few cases, occur with unerring regularity, thus behaving like exceptionally accurate clocks. Most pulsars give only weak signals so it needs large radio telescopes such as the one at Green Bank, to record the signals.

Since the effect of gravitational waves is to squeeze and stretch space-time, this will affect the arrival of the pulsar signals because the speed of light is constant.

---

3. Bell, later Dame Jocelyn Bell-Burnell, was part of the team led by Antony Hewish (1924–2021) that built and operated a novel radio telescope to investigate scintillation of radio sources at Cambridge. It was her diligence on spotting some anomalous readings that led to the discovery of the first of many pulsars by that instrument.

By comparing the arrival of the pulsar signals against atomic clocks it should be possible to detect the tiny changes in time and thus identify the gravitational waves. It is a simple concept but needs regular observations over many years before the effects can be seen. Furthermore, numerous intricate corrections need to be applied to eliminate the effects of the earth's rotation, the wobble caused by the moon, and the motions of the earth and sun. However, by June 2023 NANOGrav published firm evidence – based on 15 years of observations from 68 pulsars – that gravitational waves have been detected. Without going into details, the plot shows the observed results conforming to the prediction, the dotted line. The green line shows the expected results in the absence of gravitational waves. The period of these waves varies from years to decades, and future work will no doubt shed light on the role of gravitational waves in the early universe.

Earendel – The Earliest and Most Distant Known Star – for now?

Despite the amazing performance of the Hubble Space Telescope and the newer James Webb Space Telescope, the best images of the very earliest galaxies are little more than smudges, perhaps just a few blurry pixels in the data. With redshifts of five or more, corresponding to maybe a billion years after the Big Bang, there is no chance of resolving individual stars within such galaxies; unless, of course, gravitational lensing lends a hand.

This image, from the NASA/ESA/CSA James Webb Space Telescope, shows the massive galaxy cluster WHL0137-08, seen here alongside (at right) an inset of the most strongly magnified galaxy known in the Universe's first billion years: the Sunrise Arc. (NASA, ESA, CSA, D. Coe (AURA/STScI for ESA), Z. Levay)

Gravitational lensing is caused by the distortion of space-time by a massive galaxy or even a massive star. Given exactly the right conditions, light from a distant object behind the massive object is focussed to form a ring or arc. Christened "Einstein Rings", many have been observed, but in just a few, the lensing effect has smeared out the image so much that details of the structure can be seen, and even individual stars can be identified as tiny beads along the arc.

WHL0137-LS – or Earendel – was first discovered in HST images by a team led by Brian Welch of the Centre for Astrophysical Sciences, John Hopkins University, Baltimore. It was described as "A highly magnified star at redshift 6.2" in a paper published in *Nature* in September 2022. The work was based on a series of observations over three and a half years and verified by four independent lensing modelling processes. The spectroscopic capabilities of JWST should reveal even more details of these stars in current and ongoing studies, and will be described more fully in the next edition of the *Yearbook of Astronomy*.

Recent Advances in Solar System Exploration

Peter Rea

This article was written during the spring of 2023. As the missions mentioned are either active or due for launch imminently, the status of some missions may change after the print deadline. The mission websites are shown in bold in each section and these can be visited for the very latest information.

Introduction

The late and sadly missed astronomer Carl Sagan was instrumental in having the *Voyager 1* spacecraft take one last batch of images of our solar system, taken from its vantage point above the plane of the ecliptic. Voyager 1 took a series of images in February 1990, taking in all planets except Mercury and Mars which were too

Carl Sagan described the Earth in this image – taken from a vantage point above the plane of the ecliptic and looking down on the planets – as "a mote of dust suspended in a sunbeam". The image was captured from a distance of 5.9 billion kilometres, around 40 times the distance of the Earth to the Sun. (NASA/JPL-Caltech)

close to the Sun. In one frame, taking up less than 1 pixel of the 640,000 pixels making up the image, was the Earth, which Sagan famously named the "Pale Blue Dot". The image was taken when Voyager 1 was 5.9 billion kilometres from Earth. What this taught me was simple. Never underestimate the power of an image to inspire the next generation of young astronomers. A picture has always been worth a thousand words, and thousands of pictures from many different spacecraft and locations continue to stream towards Earth revealing the makeup of our not-so-unique system of planets and other amazing objects.

Sun: Star Explorers

The Sun is a very close star, and quite an ordinary one at that, yet it keeps our planet warm and helps plants and crops grow. It is sitting in the middle of the Hertzsprung-Russell Diagram[1] – the iconic diagram that graphically shows the life cycle of stars – where it will remain for many more billions of years. However,

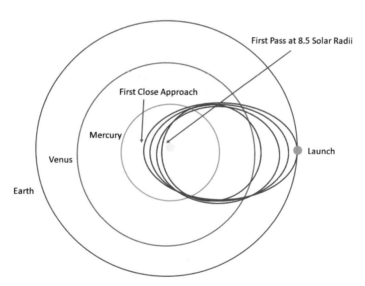

Using the gravity of Venus during repeated flybys the trajectory of the Parker Solar Probe is brought ever closer to the Sun. By the end of the mission the probe will get to within 6.9 million kilometres of the Sun. (Peter Rea)

---

1. Readers may be interested in viewing this animation of the Hertzsprung-Russell Diagram. In this video, the stars in the globular cluster Omega Centauri are rearranged according to their intrinsic brightness (vertical axis) and their temperature (horizontal axis): **esahubble.org/videos/heic1017b**

Artist's impression of ESA's Solar Orbiter spacecraft, which will observe our Sun from within the orbit of Mercury at its closest approach. (ESA/ATG medialab)

we need to keep an eye on it, as from time to time the Sun throws out intense radiation. Astronomers call such an event a coronal mass ejection (CME) and it can wreak havoc with the Earth and orbital satellites.

There are two space-based observatories, both of which are currently studying the Sun and worthy of note. The first of these is NASA's *Parker Solar Probe*, which was launched on 12 August 2018 into a highly elliptical orbit with a period of 88 days which repeatedly brings it out to the orbit of Venus. Using the gravity of Venus during a flyby, the perihelion distance – or closest approach to the Sun – can be lowered. By the time the Parker Solar Probe reaches the end of its prime mission, with its 26th closest approach to the Sun in December 2025, the perihelion distance will be 6.9 million kilometres, and that is close!

The European *Solar Orbiter* was launched by the European Space Agency on 10 February 2020 into an elliptical orbit with a period of 168 days. At perihelion it will be 0.28 au from the Sun and at aphelion at 0.91 au. The orbit is currently inclined at 24° to the plane of the ecliptic, affording good views of both the north and south polar regions. This inclination could be raised to 33° in any extended missions.

Further details and updates for both missions can be found at the following websites.

Parker Solar Probe (NASA): **parkersolarprobe.jhuapl.edu**

Solar Orbiter (ESA): **sci.esa.int/web/solar-orbiter**

Readers interested in an overview of the different aspects of the Earth-Sun environment that affect our planet may want to view NASA's Living with a Star program at: **science.nasa.gov/heliophysics/programs/living-with-a-star**

Mercury: Riding the Gravity Gradient

There is only one mission to Mercury currently en-route and that is the joint European Space Agency (ESA) and Japanese Aerospace Exploration Agency (JAXA) *BepiColombo* mission. Launched on 20 October 2018, it will take the spacecraft until December 2025 to be placed into an orbit around Mercury. As we know, Mercury has a closest approach distance to Earth comparable to that of Mars which, when at its very nearest, comes to within 54 million kilometres of our planet. However, typical Mars missions are taking between six and nine months to arrive at their goal. The predecessor to BepiColombo was the American *Mercury Surface, Space Environment, Geochemistry, and Ranging (MESSENGER)* mission, which took six

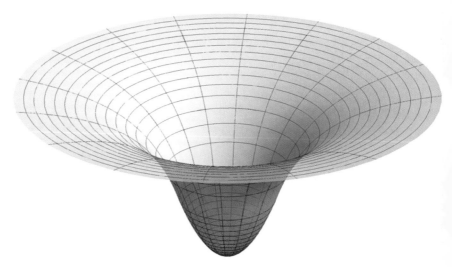

A gravity well describes a model of the gravitational field surrounding a planet or star. The more massive the body, the deeper and more extensive the gravity well associated with it. Our Sun is very massive, relative to other planets in the Solar System, so the corresponding gravity well that surrounds it appears "deeper" so planets closer to the soon have to orbit faster to stay out of the hole. (Wikimedia Commons/AllenMcC)

years and seven months to get into orbit around Mercury, a journey time similar to BepiColombo, which is taking over seven years. Why are these missions taking so long to reach their targets? The simple answer is that Mars is further from the Sun – where the Sun's gravity is weaker – but Mercury is much closer to the Sun, where solar gravity is far stronger. BepiColombo therefore must ride the gravity gradient, about which a word of explanation may be required. All objects with mass possess gravity, an apparent attractive force. Whether the object is an asteroid, a star, a

BepiColombo, Monitoring Camera 2   23 June 2022, 09:49:22 UTC

The ESA/JAXA BepiColombo mission captured this beautiful view of Mercury's rich geological landscape on 23 June 2022 as the spacecraft flew past the planet for a gravity assist manoeuvre. The image, which has north roughly towards the top right, was taken by the Mercury Transfer Module's Monitoring Camera 2, when the spacecraft was within a distance of around 920 kilometres from the surface of Mercury. Closest approach of around 200 kilometres had taken place shortly before, at 09:44 UTC. (ESA/BepiColombo/MTM)

galaxy or even an apple falling on Isaac Newton's head (though that story may be apocryphal), all possess this attractive force, and the closer we are to them the stronger is that attraction. It is perhaps best to think of the gravity field of the Sun as a hole with sloping sides. The further you go into the hole, the steeper the slope and the harder it is to get out. BepiColombo will therefore "fall" toward the inner solar system and as it gets closer to Mercury its velocity will significantly increase. This increase in velocity equates to an increase in orbital energy which needs to be dissipated for the spacecraft to be captured by Mercury. We could do this by rocket propulsion, but the velocity change needed would require very large amounts of propellant, making the spacecraft so heavy no rocket could ever launch it. If we wish to be patient, and allow a few years for the journey, we can use the gravity fields of Venus and Mercury to slowly reshape the elliptical orbit to a near circular orbit around Mercury, where the relative velocities of the spacecraft and planet are at their lowest. This method also requires far less propellant to place the spacecraft into orbit around the planet. Missions such as this may take longer, although they are more affordable. For further details, see my article "Gravity Assists – Something for Nothing" in the *Yearbook of Astronomy 2022*.

For further info on the BepiColombo mission see:
esa.int/Science_Exploration/Space_Science/BepiColombo_overview2

Venus: Back to the Morning Star

The second planet from the Sun, Venus which sometimes shines very bright as the morning star and sometimes as the evening star can get as bright as magnitude −4.6 which is bright enough to cast a shadow at a dark site. From Earth, Venus shows phases just like the Moon. It is totally covered in dense cloud, and we can only see the surface of the planet using radar imaging satellites. At optical wavelengths we only see the cloud tops. At the time of writing the Japanese have their *Akatsuki* spacecraft operating in orbit around the planet. Missions to Venus have been a little sparse of recent years, The European Space Agency launched *Venus Express* in 2005, the USA sent the radar imaging *Magellan* mission in 1989, and the last Soviet Union mission was *Vega 2* in 1985. There is, however, renewed interest in the morning (and evening) star. Rocket Lab, a USA company which has been launching small, low cost orbital rockets from New Zealand and recently from the state of Virginia in the USA, is planning to send a small atmospheric probe using their *Photon* upper stage bus. The Photon bus is based on Rocket Labs Electron upper stage. Launch would be in 2023 or 2024 and, if successful, would be the first Venus mission from a private company. Scheduled for launch in 2024

As it sped away from Venus in February 1974, NASA's Mariner 10 spacecraft captured this seemingly peaceful view of a planet the size of Earth, wrapped in a dense, global cloud layer. At optical wavelengths the surface features cannot be seem. Only radar imaging spacecraft can penetrate the cloud. However, contrary to its serene appearance, the clouded globe of Venus is a world of intense heat, crushing atmospheric pressure and clouds of corrosive acid. The morning star / evening star is not an easy planet to land on. (NASA / JPL-Caltech / Kevin M. Gill)

by the India Space Research Organisation is their *Shukrayaan-1* mission, following in the footsteps of successful Mars and lunar missions. The 2,500kg orbiter will carry an array of instruments, some of which will be home built and others with international collaboration. Missions from ESA, NASA and the Russian Federation are also planned toward the end of the 2020s but more of these in later editions of the *Yearbook of Astronomy*. On an historical note, the Soviet Union (as it was then) had far more success at landing on Venus than they did landing on Mars despite the atmosphere of Venus being acidic and very dense, giving very high surface pressures and temperatures. Mars is by far the easier of the two planets to land on and despite very low temperatures offers a more benign environment.

For further information on the Akatsuki Mission to Venus visit:
isas.jaxa.jp/en/missions/spacecraft/current/akatsuki.html

Moon: Decade of the Moon?

Could the decade of the 2020s be the decade of the Moon, just like in the 1960s? When President John F. Kennedy challenged the US nation back in 1961 ". . . to land a man on the Moon and return him safely to the Earth", nobody knew how it would be done. Although the "book" on how to do it was about to be written, there was nothing to fall back on to say how to go about it. Engineers only knew what they did not know and that was, is it safe to land? Would the lunar surface take the weight of a crewed spacecraft? Was the Moon covered in dust and, if so, how thick was it? There were more questions than answers, and before any crewed landing could be attempted those, and many other questions, had to be answered. This situation spawned the three American unmanned projects *Pioneer*, *Surveyor* and *Lunar Orbiter*. All that was over 50 years ago and today we are witnessing a return to the Moon with *Project Artemis*. Concurrent with that is a string of lunar missions from various countries or space organisations. As these words are being written we have just seen a successful flight of Artemis 1 and hopefully, by the time this volume is published, Artemis 2 carrying astronauts back to the Moon for a circumlunar flight will be close to launch. Artemis 3, taking the first woman and first person of colour to the Moon, should be around 2026. The landing will be close to the lunar south pole and some of NASA's robotic landers will be aimed toward this region. The Americans will conduct their missions under the *Commercial Lunar Payload Services* (CLPS) program, the first of which was launched in 2023 with more to follow.

Other countries are also showing interest in the Moon. South Korea launched their *Korea Pathfinder Lunar Orbiter* (officially called *Danuri*) on 4 August 2022 which achieved orbital insertion on 16 December 2022. On 11 December 2022, a Falcon 9 rocket lifted off from carrying the Japanese *Hakuto-R* lunar lander which carries the United Arab Emirates lunar rover *Rashid*. Hakuto-R will attempt a landing within the 87 kilometre diameter Atlas crater, located to the south east of Mare Frigoris. The 10 kg Rashid rover will be released by the lander shortly after touchdown. The Falcon 9 rocket also carried *Lunar Flashlight*, a tiny 14 kg NASA-developed CubeSat designed to look for water ice deposits in the south polar region to support future Artemis landings.

NASA has selected three commercial Moon landing service providers that will deliver science and technology payloads under Commercial Lunar Payload Services (CLPS). Each commercial lander will carry NASA-provided payloads that will conduct science investigations and demonstrate advanced technologies on the lunar surface, paving the way for NASA astronauts to land on the lunar surface by 2026. The one depicted here is NOVA-C from Intuitive Machines. (NASA Goddard Spaceflight Center)

This artist's impression shows the Lunar Flashlight CubeSat looking for the presence of water at the south polar region of the moon where Artemis astronauts will be sent. (NASA)

The number of future lunar missions is extensive, and space does not allow coverage of all of them here. Interested readers could put "future missions to the moon" into their preferred search engine.

Commercial Lunar Payload Services (NASA): **nasa.gov/content/commercial-lunar-payload-services**

Mars: Getting Samples Back

Since the *Yearbook of Astronomy 2024* was published, one of the Mars missions listed has ceased operations. We said goodbye to the *InSight* mission which succumbed to the ravages of Martian dust covering the solar panels, leaving InSight desperately short of power. The mission was partially successful in its science campaign. Although a device intended to burrow into the ground to measure the heat flow from the interior failed to get below the surface, a seismometer placed onto the Martian surface was very successful and recorded many marsquakes. Taking everything into account InSight was a fruitful undertaking, although nothing lasts forever.

The current Mars Sample Return mission concept envisions delivering a Mars lander near Jezero Crater, where Perseverance (far left) collects samples. A NASA-provided Sample Retrieval Lander (far right) would carry a NASA rocket (the Mars Ascent Vehicle). Perseverance would gather sample tubes it has cached on the Mars surface and transport them to the Sample Retrieval Lander, where they would then be transferred by a Sample Transfer Arm provided by ESA onto the Mars Ascent Vehicle. Also depicted is one of two Sample Recovery Helicopters that NASA will develop to be transported to Mars on the Sample Retrieval Lander, just as the Ingenuity helicopter was carried on the Perseverance rover. (NASA/ESA/JPL-Caltech)

Over the next few years NASA and the European Space Agency will develop the *Mars Sample Return* (MSR) mission. This very ambitious project seems to have been spoken of for many years, yet only now do both agencies feel confident enough that the required technology has matured sufficiently to proceed. This will be by far the most complex mission ever attempted to Mars and work is well under way. The target will be Jezero crater, where the NASA *Perseverance* rover is already collecting samples ready for collection. Ten sample tubes holding rock and regolith material have already been deposited onto the Martian surface, at a site on a flat and easily reached plain for the MSR mission to land. More sample tubes will remain onboard Perseverance and it is hoped that, if Perseverance is still operational when the MSR mission lands, the tubes inside the rover can be transferred directly to the returning spacecraft. However, because Perseverance has a lot of exploring to do between now and then, and therefore may be unable to go to the MSR lander, the tubes deposited on the surface will act as insurance that sample tubes are easy to collect. ESA was to have provided a *Fetch Rover* which would have collected the sample tubes from

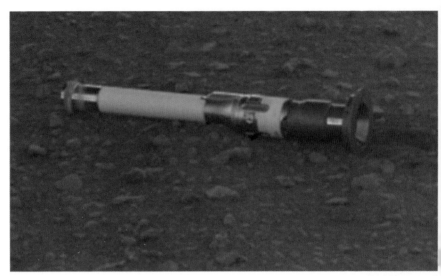

NASA's Perseverance rover deposited the first of several samples onto the Martian surface on 21 December 2022, the 653rd Martian day, or sol, of the mission. The depot will serve as a backup if Perseverance cannot deliver its samples when the Mars Sample Return mission lands at the same site. (NASA/JPL-Caltech/MSSS)

the surface and passed them to the NASA provided *Mars Ascent Vehicle*. However the fetch rover concept was abandoned in 2023 in favour of two NASA provided helicopters similar to the *Ingenuity* helicopter carried to Mars on the *Perseverance Rover*. Once samples are aboard the Mars Ascent Vehicle it will lift off from the surface to rendezvous with the ESA provided *Earth Return Orbiter* which in turn will return the capsules to Earth in a special sealed module. Current schedules put the launches of the orbiter and lander in the 2027 to 2028 timescale with an eventual return to Earth in 2033.

Perseverance rover (NASA): **mars.nasa.gov/mars2020**

Ingenuity Mars helicopter: **mars.nasa.gov/technology/helicopter**

Mars Sample Return mission (NASA): **mars.nasa.gov/msr**

Mars Sample Return mission (ESA):
esa.int/Science_Exploration/Human_and_Robotic_Exploration/Exploration/ Mars_sample_return

Asteroids: Rocky Road

There are potentially millions of small rocky bodies left over from the creation of our solar system, most of which are found orbiting the Sun between the orbits of Mars and Jupiter. Others share the same orbit as Jupiter, but occupying two of the Lagrange Points, these being L_4 – located 60° ahead of Jupiter – and L_5 – situated 60° behind. It is along this rocky road that two missions have been launched. The *Psyche* mission, part of NASA's Discovery Program, was launched in October 2023 with an arrival at the asteroid 16 Psyche scheduled for 2029. Psyche will be the first metallic asteroid visited by spacecraft.

An even rockier road awaits the *Lucy* mission, launched on 16 October 2021 on a 6.4 billion kilometre, 12 year journey to explore the asteroids co-orbiting with Jupiter. Since I reported on Lucy in the *Yearbook of Astronomy 2024*, an additional target has been added to the mission. The main-belt asteroid 152830 Dinkinesh –

This artist's illustration depicts the spacecraft of NASA's Psyche mission near the mission's target, the metal asteroid Psyche. The artwork shows the five-panel solar arrays which will not only power the spacecraft systems but the solar electric propulsion system which is used in preference to chemical propulsion thrusters. (NASA/JPL-Caltech/Arizona State University/Space Systems Loral/Peter Rubin)

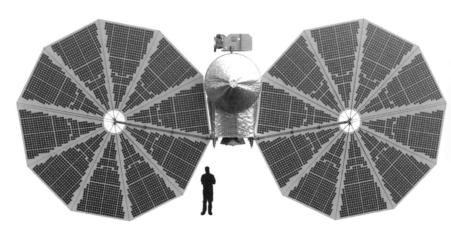

Illustration of the Lucy spacecraft with solar panels fully deployed and pictured against a human figure for scale. One of the large circular solar panels was not fully locked when deployed, although this appears to have had no effect on the mission. (NASA)

the Ethiopian name for the Lucy fossil, after which the Lucy mission is named – lies inside the asteroid belt and can now be visited with only a small course correction that will not impact on the remainder of the mission. Interested readers can keep up to date with both missions by visiting the websites listed below.

The sample return capsule from NASA's OSIRIS-REx mission, seen here shortly after touching down in the desert on Sunday 24 September 2023 at the Department of Defence's Utah Test and Training Range. (NASA/Keegan Barber)

The *OSIRIS-REx* mission to asteroid 101955 Bennu, which has been covered in the *Yearbook of Astronomy* over the past few years, has now completed its primary mission by delivering samples of Bennu back to Earth. Whilst still 101,000 kilometres from our planet the sample return capsule was spin ejected from the main spacecraft body, and on 24 September 2023 the capsule slammed into the Earth's atmosphere at 44,500 kilometres per hour. As it entered the lower atmosphere a drogue chute was deployed to stabilise the capsule, followed by the main chute deployment to allow it to make a pinpoint soft landing at the Utah Test and Training Range in the USA. Later that day the capsule was flown to the Johnson Space Centre outside of Houston in Texas where it would be opened to reveal the samples that will keep scientists busy for many years. Shortly after the return capsule had been ejected the spacecraft thrusters fired to accelerate it away from Earth in order that it can continue on an extended mission, now renamed *Osiris Apex* (Apophis Explorer). Its next destination is the asteroid 99942 Apophis, where it should arrive in April 2029.

Psyche (NASA): **psyche.asu.edu**

Lucy (NASA): **lucy.swri.edu**

OSIRIS-REx (NASA): **asteroidmission.org**

Jupiter: Exploring the Icy Moons

First imaged from the *Pioneer 10* and *Pioneer 11* spacecraft, the Galilean moons of Jupiter were observed up close by the *Voyager 1* and *Voyager 2* spacecraft in 1979, the images and results returned showing fascinating worlds worthy of further investigation. The *Galileo* spacecraft studied the planet Jupiter and its satellites for almost eight years following its arrival at Jupiter in December 1995. The *Cassini* mission in 2000 and *New Horizons* in 2007 observed them during brief flybys. It has long been speculated that water oceans lie beneath the surfaces of both Europa and Ganymede, and it is time to go back and explore these moons. The European Space Agency has recently launched the *Jupiter Icy Moons Explorer* (JUICE) mission which will arrive in orbit around Jupiter in July 2031. Following in the footsteps of JUICE is NASA's *Europa Clipper*, scheduled for launch in October 2024 for a planned arrival at Jupiter in April 2030. Interested readers can keep up to date with both missions by visiting the websites listed below.

Juice Mission to Galilean Moons (ESA): **sci.esa.int/web/juice**
Europa Clipper (NASA): **europa.nasa.gov**

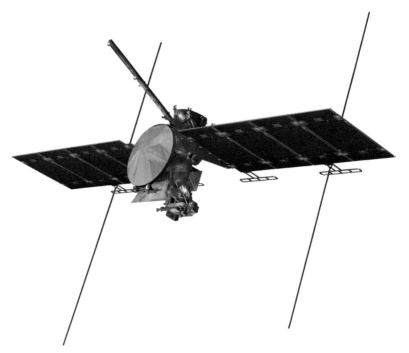

Artist's rendering of the Europa Clipper spacecraft due to launch toward Jupiter and its moon Europa in 2024. (NASA/JPL)

Other useful Solar System exploration websites include:

NASA Mars Program: **mars.nasa.gov**

NASA Solar System Exploration: **solarsystem.nasa.gov**

ESA solar and planetary mission operations: **esa.int/Enabling_Support/ Operations/Solar_system**

As always, Solar System exploration continues to excite and inspire, and next year promises to be no different.

Anniversaries in 2025

Neil Haggath

Johann Bayer (1572–1625)

We are all familiar with the system of naming stars in each constellation with Greek letters. This was established by a German amateur who died four centuries ago.

Johann Bayer was born in Rain, Bavaria in 1572. He was a lawyer by profession, and an amateur astronomer. Almost nothing is known about his life or personality.

In 1603, Bayer published his famous star atlas, *Uranometria*, which has been called the greatest contribution to astronomy by an amateur. It depicted, with both accurate star mapping and beautiful illustrations of the constellation figures, all the 48 traditional constellations of Ptolemy's *Almagest*, plus 11 new ones in the region around the South Celestial Pole, which is never seen from Europe. The latter were invented by the Dutch-Flemish astronomer and cartographer Petrus Plancius (1552–1622) five years earlier, but Bayer's inclusion of them led to them becoming widely recognized; they are now among the 88 constellations officially named by the International Astronomical Union. *Uranometria* became very popular, and was the most widely used star atlas during the seventeenth century.

Until then, there had never been any systematic method of naming stars, other than the traditional Greek or Arabic names of the brighter ones. Bayer introduced a system whereby the stars within each constellation were named with Greek letters, followed by the Latin genitive of the constellation name. Usually, they were lettered in order of brightness; the brightest star in Taurus is Alpha (α) Tauri, or "Alpha of Taurus", the second brightest is Beta (β) Tauri, and so on, although there are some errors; for example, β Orionis (Rigel) is brighter than α (Betelgeuse).

There are some notable exceptions, where the system was not followed; for example, the seven brightest stars of Ursa Major (The Great Bear) are lettered in order of their positions (from east to west) in the Plough or Big Dipper asterism.

Despite its strange idea of combining Greek and Latin, Bayer's system of nomenclature caught on, and we still use it today. But it has an obvious limitation – the Greek alphabet only has 24 letters!

John Flamsteed (1646–1720) later used a system using numbers within each constellation, which therefore had no limit; he assigned numbers to almost all the stars visible to the naked eye. The convention in modern star atlases is to

Orion as depicted in Bayer's *Uranometria*. (US Naval Observatory Library)

use Bayer's letters for those stars which have them, and Flamsteed numbers for the rest.

Bayer became involved in another project which was far less successful. Another German amateur, Julius Schiller (c.1580–1627), decided to replace all the "pagan" constellation names with Christian ones, and renamed them all after Biblical figures and religious symbols. He somehow persuaded Bayer to help him produce an entire new star atlas, the *Coelum Stellatum Christianum* which, when it was published, was almost entirely ignored.

Johann Bayer died in Augsburg on 7 March 1625.

The Speed of Light

This year marks the 350th anniversary of an important "first" – the first determination of the speed of light, by Danish astronomer Ole Rømer in 1675. This is described in the author's article 'The Speed of Light', elsewhere in this volume.

The Royal Observatory

In the seventeenth century, seafaring nations needed to find a reliable method of determining longitude at sea for navigation. The method proposed involved measuring the changing position of the Moon against the stars; this required more accurate star charts than any yet produced.

The Royal Observatory at Greenwich was established for the purpose. It was commissioned by King Charles II; the foundation stone was laid on 10 August 1675, and the observatory completed the following year. It was built in Greenwich Park, which was a Royal estate. The main building, with the famous Octagon Room, was designed by Sir Christopher Wren.

Rev. John Flamsteed (1646–1720) was appointed as the first Director, and given the title of Astronomer Royal by the King – although he had to provide his own telescopes! He held the post until his death; the main observatory building was later named Flamsteed House.

Flamsteed produced a star catalogue that was far more accurate than any before, though it took many years. Within a few decades, the development of accurate chronometers rendered the "lunar distances" method obsolete, but the observatory remained an important centre for positional astronomy and timekeeping. *The Nautical Almanac*, founded in 1767, is still published today.

In 1851, the longitude of Greenwich was internationally adopted as the Prime Meridian, the zero line of longitude, and Greenwich Mean Time as the worldwide standard for timekeeping. The actual meridian is defined by the Airy Transit Circle, installed by Sir George Airy.

The Old Royal Observatory in Greenwich Park. (Wikimedia Commons/Kjetil Bjørnsrud/ GFDL 1.2 or later)

In 1957, the Royal Observatory was relocated to Herstmonceux Castle in Sussex, and again to Cambridge in 1990. The original observatory is preserved as a museum.

Until 1972, the Observatory's Director always held the title of Astronomer Royal. Thereafter, the latter became a purely honorary title, which was awarded to eminent researchers from other fields of astronomy.

The Unluckiest Astronomer

Three centuries ago – on 12 September 1725 at Coutances, France – a man was born who is often described as the unluckiest astronomer who ever lived. He was the bizarrely named Guillaume-Joseph-Hyacinthe-Jean-Baptiste le Gentil de la Galaisière, perhaps best known for his epic voyage to India, to fail to observe both the 1761 and 1769 transits of Venus. I plan to tell his tragicomic story in a future edition of the *Yearbook of Astronomy*.

Four Deaths in 1875

The year 1875 saw the deaths of four notable astronomers, three German and one English.

Friedrich Wilhelm August Argelander (1799–1875) founded the Bonn Observatory in 1836. In 1863, he published the *Bonner Durchmusterung*, a catalogue of the Northern Hemisphere of the sky listing the positions of over 324,000 stars down to magnitude 9.5. Later extended by others to cover the southern sky, this remains a standard work.

Heinrich Louis d'Arrest (1822–1875) worked at Berlin and later Copenhagen Observatories. On 13 September 1846, while an assistant at Berlin, he and Johann Galle searched for and discovered Neptune, following Urbain Jean Joseph Le Verrier's prediction of its existence.

Heinrich Schwabe (1789–1875) was an amateur astronomer, remembered for his work on sunspots and his discovery of the 11-year sunspot cycle.

English amateur Richard Christopher Carrington (1826–1875) was another eminent observer of the Sun. Among many contributions, he made the first observation of a solar flare, and independently discovered Spörer's Law, which concerns the variation in latitude of sunspots.

Camille Flammarion (1842–1925)

One of the most prolific astronomy and popular science writers of his time died 200 years ago.

Nicolas Camille Flammarion was born on 26 February 1842 in Montigny-le-Roi, France. At 16, he began work as a computer at the Paris Observatory.

From an early age, he speculated about life on other planets, and published his first and probably best known book *La Pluralité des Mondes Habités* (The Plurality of Inhabited Worlds) in 1862. Later the same year, he was sacked from the Paris Observatory; it is not known whether the two events were related!

Camille Flammarion. (John McCue)

Flammarion published over fifty books, mostly on astronomy and popular science, but including some science fiction novels. From 1882, he published *L'Astronomie* magazine.

In 1887, he founded the *Société Astronomique de France* (SAF), and became its first President. In 1883, he established a private observatory in Juvisy-sur-Orge; this is now owned by the SAF, and operated as a public observatory.

Flammarion also had some unconventional ideas. He believed in Percival Lowell's "Martian canals", and researched ghosts and hauntings, and other supposed "psychic" phenomena. He was widely misquoted in the press as fearing that the Earth's passage through the tail of Halley's Comet in 1910 might poison all of humanity, although he had in fact responded to such ideas by saying that it was "improbable", and correctly saying that, as the gas in the tail is extremely tenuous, it would have no noticeable effect.

Flammarion was made *Commandeur de la Legion d'Honneur*, France's highest order of merit. He died in Juvisy on 3 June 1925, aged 83.

Handshake in Orbit

Fifty years ago, the mood of manned spaceflight briefly changed from competitive to cooperative, with the first joint mission between the United States and the Soviet Union – the Apollo-Soyuz Test Project (ASTP).

With political relations between the two powers improving, an agreement had been signed three years earlier, to conduct a joint mission in which an American Apollo spacecraft would dock with a Soviet Soyuz, and the crews would visit each other's spacecraft. As well as being a symbolic show of political cooperation, it was intended to develop a mutual rescue capability, in case of any future space accident.

But docking an Apollo with a Soyuz was far from straightforward, as their docking mechanisms were not compatible. So a special Docking Module was built, which would be carried into orbit by the bigger Apollo; one end of it had a standard Apollo docking mechanism, and the other a mechanism built to be compatible with that of the Soyuz. The latter was designed to be "androgynous", enabling either spacecraft to perform the active or "male" role – though the more versatile Apollo would perform the delicate rendezvous manoeuvres.

Another problem was that, while Apollo spacecraft had always been pressurized with pure oxygen at only five pounds per square inch, Soyuz used normal air. So had the Soyuz crew gone directly into the Apollo, they would have suffered the bends! The Docking Module, therefore, also had to function as a decompression chamber. To make the process easier, this Soyuz used a reduced pressure, oxygen-enriched atmosphere.

The ASTP crews, (left to right) Deke Slayton, Tom Stafford, Vance Brand, Alexei Leonov, Valery Kubasov. (Johnson Space Center/NASA/Photo ID: S75-22410)

The two crews, and their backups, were assigned two years ahead, and their intensive training included visiting each other's countries and learning each other's languages.

The American crew was commanded by Brigadier-General Tom Stafford USAF, a veteran of three previous spaceflights. His colleagues, Vance Brand and Donald "Deke" Slayton, were both rookies. Slayton, the Chief of the Astronaut Corps, was one of the "Original Seven" astronauts selected for Project Mercury in 1959, but was grounded due to a medical problem; he waited 16 years for his first and only spaceflight at the age of 51!

The Soviet commander was Air Force Colonel Alexei Leonov, who had made the first spacewalk ten years earlier. His colleague, Valery Kubasov, was also making his second flight.

Both spacecraft were launched on 15 July 1975 – Soyuz 19 first, and the unnumbered Apollo craft 7½ hours later. Fifty two hours into the mission, the two

A mockup of the docked ASTP spacecraft displayed in the Museum of the Energia Rocket and Space Corporation near Moscow. Below it is the actual Soyuz 19 crew module. (Neil Haggath)

docked. In the Docking Module, Stafford and Leonov shook hands and greeted each other in each other's languages. They remained docked for two days, with each crew visiting the other's spacecraft and performing joint experiments.

After the spacecraft undocked, they briefly docked again, this time with the Soyuz taking the active role. Then they finally parted, and Soyuz 19 returned to Earth after just under six days. Apollo remained in orbit for three more days, with a busy scientific programme; it would be America's last manned spaceflight for six years, before the first space shuttle flight.

The Apollo crew almost met with disaster on landing. The parachutes deployed late, causing the capsule to hit the water faster than normal and toxic gas from a thruster leaked into the cabin. Thankfully, the crew escaped unharmed.

Before long, political relations had soured again, and the Soviet Union began a programme in which all its communist allies supplied guest cosmonauts – including a Vietnamese Air Force officer who claimed to have shot down an American bomber. There would be no further Russian-American cooperation in space until after the collapse of the Soviet Union in 1991.

Skies over Ancient America
Mystical Mounds and Landmarks of the Prehistoric Americas

P. Clay Sherrod

Following on from Part 2 of 'Skies over Ancient America', published in the *Yearbook of Astronomy 2024*, this final section completes the series of articles covering this fascinating topic, and deals with great efforts of construction – both elaborate and simple – that appear at first examination to provide either observations of or predictions for important calendrical dates throughout the year. The forecasting of such annual dates became more and more critical to the earliest cultures of the Western Hemisphere, as their populations increased along with the need for seasonal essentials such as water the planting and harvesting of their growing dependence on crop plants. Society was rapidly moving from the hunter-gatherer concept of survival to that of planned and sophisticated farming. We also examine the curious representations of western cultures of various aspects of the nighttime sky – constellations, the Milky Way, and sky events – that indicate the importance in perhaps a spiritual reverence for the abyss of space.

Similarly, stone and earthen mounds throughout Mesoamerica and farther north into the areas of what is now the United States and Canada likely were used for solsticial and equinox predictions as demonstrated with over 33 mound sites progressing from the states of Louisiana into northern Wisconsin in the study carried out by Sherrod and Rollingson.

The culture that existed in the northern Americas since as early as 5000 BCE seems to have progressed and spread northward from Louisiana, likely following the amelioration of climate the life-supporting waters of the Mississippi River. This culture of people – the "mound builders" – is poorly understood and little is known about their ethnicity although many scholars have attempted to associate them with their Mesoamerican counterparts. In early archaeological circles, these people were referred as "The Toltecs" and indeed the mound building and community centre layouts do indeed closely match that culture. Nonetheless there is no evidence to link the North American Mound Builders to southern Mesoamerican cultures.

The Aztec god Quetzalcoatl, as depicted in the sixteenth-century collection of Aztec manuscript painting known as the *Codex Telleriano-Remensis*. Quetzalcoatl was a deity recognized in the earliest Aztec culture and religious worship. Among the Aztecs and later into subsequent cultures, it was important to wind, Venus, Sun, trading, arts, crafts, knowledge and learning, and its widespread power led to this creature becoming the patron god of the Aztec priests and rulers. More information about the Aztecs and their culture appears later on in this article. (Wikimedia Commons/Artist Unknown)

Not only were the largest mound sites in North America possibly aligned with the rising and setting of the sun during critical times during the year, but the research reveals what is now known as the "Toltec module", a measuring increment of both mount sizes and spacing of 47.5 metres, 95 metres, 190 metres and every multiple of 47.5. It is shown that the number 1.727 metre is the only number that will divide into this module and that value just happens to be the average height of the adult male – perhaps a priest – some 2,000 years before present. In nearly all northern American mound sites, layers of earth were added periodically to the height of mounds to an average of 1.727 metres.

Attempting to match this "module" in the pre-Columbian Mesoamerican sites farther south, such as Teotihuacan and Chitch'en Itza, was unsuccessful but very strong indications of a measurement module of 52 metres was evident throughout the Mayan and other Mesoamerican developments.

Mound building was an important part of community life and worship, in addition to adding seasonal prediction capability, in cultures from Central America into the Northern America regions as migrations relocated ancient traditions and peoples following the life-sustaining Mississippi River upstream to the present Canadian border. The following selection covers only a few of the hundreds of mound sites from Central America stretching northward into what is now the northern United States.

Watson Brake – Louisiana

Outside of Mexico and Central America, what is believed to be the oldest mound site in North America is Watson Brake, in northern Louisiana. Excavations have revealed that this early culture began constructing a group of 11 mounds in an oval arrangement 5,400 years ago. The long axis of the oval mound complex is curiously aligned to the winter and summer solstices. Numerous artefacts found at the site, including projectile points and earthenware, have helped in dating the occupation of the mound site, but there is no evidence that clearly points to specific ceremonial or time-keeping functions.[1]

Serpent Mound – Ohio

In Ohio the 411 metre long Serpent Mound is an extraordinary earthen effigy that obviously required much concentration, time and resources to build; the serpent is the largest effigy mound (depiction of an object or animal) in the world, the representation being the multi-curved writhing of a great snake. The multi-layer construction is sophisticated, and the head of the serpent appears to be directly aligned with the setting sun on the summer solstice. Serpent Mound was built

1. Romaine, William F., 'Ancient Skywatchers of the Mississippi Valley. Power Point presentation for The Maya Exploration Center', 27 September 2019' academia.edu/40585226/Ancient_Skywatchers_of_the_Mississippi_Valley

at least 1,000 years ago by the Adena culture, non-tribal except to be known as "the mound builders." More information on Serpent Mound can be found at **ohiohistory.org/visit/browse-historical-sites/serpent-mound**

Cahokia Mounds – Illinois

The construction of the largest mound complex known to have existed – as well as the most populated of all mound ceremonial centres – started at about 700 CE by what appears to be roughly 1,000 people known as the "Mound Builders." Cahokia Mounds has been decimated by modern construction and excavation, but at one time contained more than 120 mounds, some the largest earthen mounds in the world. Although most are now gone, still remaining are spectacular representations of the dedication and labour involved in building this enormous ceremonial and residential site. For further details visit: **cahokiamounds.org**

At its peak, this unknown culture reached, at about 1100 BCE, an astonishing population of perhaps 150,000, by which time starvation and almost total destruction of the environment and ecosystem had occurred. The entire mound complex covers an area of a little over 15 square kilometres. Construction on the largest mound – Monk's Mound (so named for the French Trappist Monks who lived on and around the mound starting in 1809) – began around 900 CE and continued getting larger and taller until what were likely desperate times in 1200.

From a small terrace known as "Fowler's Point" about halfway up the southwest corner of Monk's Mound there may have been a religious observatory vantage point earmarked by a large cedar pole that was continually replaced as the huge mound was expanded and layers were being added. From this vantage point a very clear view of nearly the entire site afforded somewhat of an engineer's "transit" from which the site is laid out in the cardinal directions of North, South, East and West, and nearly all mounds

The Cahokia Birdman Tablet: effigy figures and clay tablets have been recovered at Cahokia that are intriguingly similar to others at the Mexican city of Teotihuacan (see following) and other astronomically significant pre-Columbian sites. (Wikimedia Commons/Herb Roe/chromesun.com)

were spaced in accordance with the Toltec Module (multiples of 47.5 metres), and their construction dimensions also of that module.[2]

Mound alignments to solstice and equinox positions dominate the Cahokia site; perhaps most notable among them is Mound 72, a unique burial mound for one individual of great importance either to the Cahokia culture or more possibly to the entire mound building culture of North America at that time. This individual was ceremonially placed in a very unimpressive long and oval earthen mound that is oriented to the winter solstice sunrise and summer solstice sunset. The mound itself is located next to the location of what was a huge cedar post which aligns to north-south to the post previously mentioned at Fowler's Point, again in increments of the Toltec Module. Laid upon a bed of 20,000 seashells carved into beads and arranged in the shape of a half-bird/half-human, this person of great importance was buried at about the age of 40. The burial was macabre to say the least: in what was obviously a sacrificial offering, 53 young women were strangled and buried alongside in a separate mound, as were 39 "soldiers" who had been either savagely bludgeoned or buried alive adjacent to Mound 72. Offerings from locations possibly thousands of kilometres distant accompany this burial, indicating strong evidence of trade as well as the high significance of this individual in life.

Toltec Mounds – Arkansas

The mound builders of the Arkansas site of Toltec Mounds were originally erroneously named "The Toltec Mounds" by the landowner of the mound site Mary Elizabeth Knapp, as her belief was that the earthworks resembled those of the Mexican Toltec culture. For further details of the Toltec Mounds Site visit: **encyclopediaofarkansas.net/entries/toltec-mounds-site-413**

Thought to have been abandoned around 1050 CE more than a dozen large earthen mounds were enclosed by a wall, or "embankment" that likely was to protect the site from flooding by the nearby oxbow lake (Mound Lake) by the nearby Arkansas River. Some occupation at the site suggests earlier activity, but the Toltec Mounds were built and used likely for a 300-year period the site was occupied. Eighteen mounds were arranged around two rectangular open spaces that were used for ceremonies. Most of the mounds have been destroyed by modern agriculture today, but one mound remains at 15 metres high, and a flat "terrace mound" is around 12 metres high, while a third mound which has been found to be heavily laden with midden, is still a little over four metres high; the

---

2. Sherrod and Rollingson, Surveyors of the Prehistoric Mississippi Valley, *Arkansas Archeological Survey Research Series*, **28**, 1987.

Thought to have been constructed over a thousand years ago, the huge earthen Mound A at Toltec Mounds archeological site towers behind the smaller priests' mound, Mound B. Both of these huge earthen structures – similar to arrangements in dozens of others in the Mississippi River Valley – appear to serve as solsticial markers for the rising and setting positions of the sun on the horizons during either winter or summer solsticial dates." (Wikimedia Commons / Herb Roe / GNU-FDL / License CC BY-SA 3.0)

original heights are unknown. The others that remain are less than 1.2 metres high. Only one mound has been identified as a burial mound, likely a priest or high figure. The three largest, and remaining, mounds were possibly used as platforms for ceremonies and those align very well with solar azimuths for the solstices, while the terrace mound is thought to have had residences of religious leaders on them. Several mounds were positioned to line up with the sun on the horizon at sunrise The standardized unit of measurement, 47.5 metres, was extensively used at Toltec and it was there that the "Toltec Module" was first suspected by Sherrod and Rollingson in the placement of mounds, so that the site layout was not happenstance with even the open plaza centred within the mounds having an overall size based on that module.

Mesoamerican Structures (Southern)

Nearly all of the southern Mesoamerican cultural sites predate the North American mound sites. Although there are striking similarities in the construction of mounds at some sites – such as Cahokia – there appears to be little evidence to support an actual ethnic or commerce-relate link between southern and northern sites of prehistory.

Large structures dominated the landscape in Mesoamerica but these differ from the North American mound sites in that the structures were built of earth and then a façade of large stone was sculpted and laid over the mound to preserve its shape in all Mesoamerican sites. Absolutely no stone covering was used in any of the American sites, but it must be noted that the chronological first migration from Central America into North America may have been the Louisiana sites, a river delta area in which very little stone is available.[3]

Olmec – Origin of Mesoamerican Communities?

Perhaps the oldest known culture of the ancient Americas is that of the Olmec civilization which pre-dated the Aztecs and Mayans by as much as a thousand years; very little is known of the Olmec people and their way of life outside of the huge monuments – mostly enormous stone heads carved deeply into native boulders – that were left behind.

The Olmecs settled into a small area near the present day Yucatan Peninsula (Veracruz and Tabasco) near the shores of the Gulf of Mexico. Starting their communities perhaps as early as 1200 BCE farther south that either the Mayan or the Aztecs, the Olmec people were dispersed or assimilated by the people of other cultures by 400 BCE.

Matthew Williams Stirling, former director of the Bureau of American Ethnology, shown with the enormous Olmec head, received from the National Museum in Mexico City, at the exhibit at Museum of Natural History. This remarkable stone carving of unknown purpose stands as the most significant remainder of the ancient Olmecs, the earliest known major Mesoamerican civilization. (Courtesy of Smithsonian Institution Archives)

---

3. Further information can be found at the Poverty Point National World Heritage Site **povertypoint.us**.

Very little is known about the Olmec, much less than we have been left by the Aztec and Mayan cultures; indeed there is a significant overlap in what previously was thought to be early Mayan artefacts that are now know to be Olmec. In fact, there is some suggestion, although little is in the scientific treasure chest, that the Olmec people had their roots in Africa.

Although they had developed very sophisticated water transfer and drainage systems throughout their enormous urban centre of San Lorenzo Tenochtitlan, there is no evidence of any importance to the night sky, no definite alignments to be found and no calendar markers or stones as we find in nearly all other Mesoamerican cultures. Nonetheless, we do know for certain that the Olmec technologies and infrastructural methods were adopted by the later Mayan and Aztec rulers, thus the influence of the Olmec to upcoming cultures which would flourish later was great. For further information visit: **worldhistory.org/Olmec_Civilization**

Maya

There is no doubt that the use of astronomical alignments and the importance of celestial objects such as the sun, moon and Venus were of high and religious importance to the Mesoamerican people, particularly the Mayan. The literature is filled with studies showing an array of solar, lunar, and planetary alignments, and a relative abundance of artefacts and written records offer insight into the cultural uses of cosmological knowledge. Stone carvings, paintings, bark-paper books, and descriptions written by early Spanish invaders have all provided important insights; unfortunately, those invaders destroyed much of the keys to understanding the exact relationship of sky, earth and human in the Mayan culture. For further information on the Mayan civilization visit: **history.com/topics/ancient-americas/maya**

Teotihuacan

In relation to other Mesoamerican cultures Teotihuacan was contemporary with the early Classic Maya (250–900 CE) but earlier than the Toltec civilization (900–1150 CE).

Predating Teotihuacan was the extraordinary city of Coba which was occupied by up to 50,000 people as early as the last century BCE, and the residence of which appear to have had numerous conflicts and disputes with those of the later Teotihuacan. Located near present day Mexico City, the namesake of Teotihuacan is curious as it is not Mayan, but Toltec in origin, leading speculation that the enormous structures of the more recent city were built atop more ancient temples of the Toltec culture.

The most recent city was constructed between 150 BCE and 200 CE – almost contemporaneous with Coba – and benefited from a plentiful supply of spring water which was channelled through irrigation. The largest structures at the site were completed before the third century CE, and the city reached its peak in the fourth century CE with a population as high as 200,000. Teotihuacan is actually the Aztec name for the city, meaning "Place of the Gods". Other deities often represented in Teotihuacan art and architecture include the feathered-serpent god known to the Aztecs as *Quetzalcoatl*, who represented agricultural renewal (especially maize), and the creator god known as the Old Fire God. The positioning of temples and pyramids in alignment with the sun on the June solstice and the Pleiades suggests calendar dates were important in rituals, and the presence of buried offerings and sacrificial victims illustrates the belief in the necessity to appease various gods, especially those associated with climate and fertility.

Archaeoastronomers Victoria and Harvey Bricker, using the list in the Paris Codex's animal known significance, have compiled a number of possible constellations and star associations as perhaps recognized by the earliest Mayan people. The following list demonstrates at least a portion of the constellations that the ancient Maya may have so designated.[4]

Sipakna – the mover of the sky and plays with the volcanoes, and is stretched across the sky as the Milky Way.

Xik – means "hawk" in K'iche' Mayan, and the present day stars that form Pisces and Aquarius.

Crux Eloq'omab' – this constellation refers to what is commonly known as the Southern Cross.

Ixtutz'/Ixpeq – today Capricornus, this represented the legs of a toad as served ceremonially as meals during ceremonies for rainfall.

Jun Raqan Tukur – means "one-legged owl are our present day stars of Virgo and Libra.

---

4. For more about ancient Maya life and, in particular, constellations and the Ancient Maya, visit: **ancientmayalife.blogspot.com/2018/02/constellations-and-ancient-maya.html**

Korkoche – the modern day outline of our canine constellation Canis Major, this constellation does NOT represent an animal, but the sound of one: the Woodpecker!

Kej – which means "deer" in Mayan and was possibly our present constellation of Sagittarius.

Pa Raqan Kej – an interesting Mayan name of our constellation Scorpius which is actually completed as "scorpion biting under the leg of the deer".

Oxib' Xk'ub' – this star designation was the three bright stars of Orion's belt that we envision today, to the Mayans represented three stones protecting their fires.

Aztec

Further north from the dominant Mayan influence, in what is now Mexico, scientists have studied how the Aztec culture incorporated calendars and cosmological happenings into the culture of otherwise ordinary life. Like the Maya, the Aztecs adopted interlocking calendars based on 360- and 260-day cycles, with ceremonial days linked to the movements of the sun, moon, and stars. For further information on the Aztec people and their culture visit: **en.wikipedia.org/wiki/Aztecs**

As with the Maya influence, there is evidence that the architectural alignment of Aztec cities, including their capital Tenochtitlan, highlight the seasonal movements of the sun and stars. As American anthropologist and astronomer Anthony Aveni notes, such arrangements, "… sealed the workings of the cosmos into the architectural fabric and sacred space of the city."

The Aztec Calendar Stone was discovered by the Spanish during the first modern construction in Mexico City in 179 CE. The Spanish-American astronomer and polymath Antonio de León y Gama (1735–1802) became the first scholar to try to understand the astronomical system represented in the calendar stone, this apparently having been carved with intricate cosmological motifs somewhere around 1500 CE.

The sculpted motifs that cover the surface of the stone refer to central components of the Aztec beliefs of their people as they were bonded to the cosmological spectrum of all of creation. The sacred stone demonstrates via the intertwining of the motifs, the concepts of Aztec ideology such as the ever-dominant realm of violence and warfare of their era, the cosmic cycles, and the nature of the relationship between their deities and man. It was a powerful tool; the Aztec rulers used this association with the cosmos and the bloodshed often associated with their lives, and depicted through the Sun Stone, to maintain control

The Aztec Sun Stone (or Calendar Stone) depicts the five consecutive worlds of the sun from Aztec mythology. This very complicated carving may – or may not be – a true calendar, but rather a depiction of Aztec worship. The stone may be a functioning calendar, but rather it is an elaborately carved solar disk, which for the Aztecs and other Mesoamerican cultures represented rulership. At the top of the stone is a date glyph (13 reed) which represents both the beginning of the (then present) sun, the 5th and final one according to mythology, and the actual date 1427 CE, thereby legitimizing the rule of Itzcoatl (who took power in that year) and creating a bond between the divine and mankind. Whatever the purpose, this remarkable work of architecture and geometry by the Aztecs does consist of a way of calculating time. It consists of a solar and a sacred calendar. The solar calendar was used to calculate the 365 days of each year. The sacred calendar consisted of 260 days. Both these calendars coincided after every 52 years which formed one Mesoamerican century and this is depicted clearly with symbols at the appropriate places on the stone. Roughly every century the New Fire Ceremony was held during which human sacrifice and other rituals were offered. (Wikimedia Commons/Juan Carlos Fonseca Mata/ License CC BY-SA 4.0)

over the population. This intricate carving was a tool the Aztec leaders' used as a method of citizen control over their people, all in the name of astronomical events in the skies above.

Once Antonio de Leon y Gama had studied the mysterious stone in depth, he argued that only Mexicans could properly understand Aztec culture, as Mexico embraced both the European and Aztec heritage. By 1800, he was convinced that the Aztecs had been on the same cultural level as the great civilizations of ancient Greece and Egypt.

Inca

Perhaps more so than with any other western prehistoric culture, the Incan people observed and worshipped the sky from both a cosmological and religious standpoint; their devotion to it is clear in the many tantalizing archaeological sites that were left behind. To the Incan, the expanse of the sky, the Earth, its people and all that were contained within it were intertwined. Since domestic agriculture was practiced, the need for calendars was also present. Further information on the Inca civilization can be found at: **newworldencyclopedia. org/entry/Inca_Civilization**

The Incans were able to imagine and outline constellations, bright stars of importance, the passage of short and long intervals of time, predict the seasons and more, all from their study of the sky. The premier Incan cultural centre – Machu Picchu – was designed and built according to the deity of the sun and the seasonal motions of the sun. Other cities also demonstrated astronomy as an important aspect of life; the decaying city of Cusco is lined with streets that depict the constellations inscribed within them.

As with other contemporaneous cultures, the Incas worshiped celestial gods: *Viracocha* (The Creator), *Inti* (the Sun), the stars and *Mama Killa* (the Moon). However, differing from other cultures and somewhat mirroring the beliefs of the North American Hopi, the Incan people held as gods the mountains, the rivers, the Earth and other aspects of the natural world.

Like the ancient Egyptians, the Inca culture was based on azimuth positions of celestial objects on the horizon. They built carefully placed pillars and carved monoliths on mountains and hills which allowed them to identify the passage of time with great accuracy.

Inca Constellations: Light and Darkness

The Incas specified many "constellations", as resembling what they saw in their daily lives (animals, human activity, fire, water). They believed that Viracocha had

The magnificent Pleiades – a star pattern of influence throughout antiquity – drawn by the Mayans as a rattlesnake. Aries was drawn as an ocelot; Gemini drawn as a bird they called a cox bird or bird 2; Pisces drawn as a skeleton; Scorpius drawn as a scorpion; Aquarius drawn as a bat; Libra drawn as an animal they called bird 1; Sagittarius drawn as a fish-snake; Capricornus drawn as an animal they called bird 3; Virgo drawn as an animal they couldn't identify; Orion drawn as a turtle. (Arkansas Sky Observatories / P. Clay Sherrod)

given every animal, bird and every living thing a corresponding star that protected it and thus associating the stars into constellations for these gifts became a very important practice for the Inca civilization.

The importance to animals in the Incan world becomes very evident when we learn that the famous star cluster – the Pleiades – one of the most conspicuous asterism in the night sky was first and foremost directed to serve as representative of the god that looked over the welfare of all animals.

The second type of religiously important sky consisted of the dark clouds of the Milky Way. These dark areas (dark dust as we know them today) must have been awe-inspiring in the pre-light-polluted era of Mesoamerica. The dark lanes were considered animate constellations (living animals) and the Milky Way was considered an enormous river, much as it was in Egypt almost halfway around the world. It was this river that was traversed by all spirits that finished their journeys on Earth. The Incas are one of the few ancient civilizations that were able to identify constellations without the presence of stars, using instead the contour of the light and dark portions of the Milky Way to perceive figures of the gods that they held in esteem.

The Nazca Lines

Created possibly as long ago as 500 BCE, the Nazca Lines are famous both for their incredible intricacy as well as the complexity of rendering such large linear figures over hills and valleys while maintaining the true linear boundaries that are exhibited within each of these huge motifs. For more information on the Nazca Lines visit: **history.com/topics/south-america/nazca-lines**

Attempts have been made to associate these curious features – some of which are nearly 50 kilometres long – with celestial explanations. The figures can only be clearly identified when viewed from high above. There is a curious assortment including humanoid, animal, and even geometric, the latter leading to speculation about alien spacecraft runways; but thus far only conjecture rules. It is possible that the many of the more-than 300 figures represent the Incan interpretations of the constellations of the night sky – in addition to long, linear and geometric symbols throughout the Andes Mountains, there are representations of a spider, hummingbird, cactus plant, monkey, whale, llama, duck, flower, tree, lizard and dog. Interestingly, one huge outline is nicknamed "The Astronaut" from an uncanny resemblance to a twentieth century human in a space suit.

Assessing Ancient Efforts in American Skywatching

We are at a tremendous disadvantage in understanding the true association with carvings, paintings, earthen and stone mounds, and true intent of the ancient American cultures in regard to most of the symbols and structures that we believe might be astronomical in intent. For example, is Cahokia's Woodhenge truly a sophisticated circle of posts, or merely utility poles that modern researchers have

Spanning an incredible 93 metres by 58 metres, and with its huge and distinctive spiralling tail, the Nazca "Monkey" appears from space to be exactly that. When viewed from ground level, this incredible feat of engineering cannot even be discerned as anything other than a very long and continuously meandering line of pile stones. (Wikimedia Commons / Diego Delso / delso.photo / License CC BY-SA)

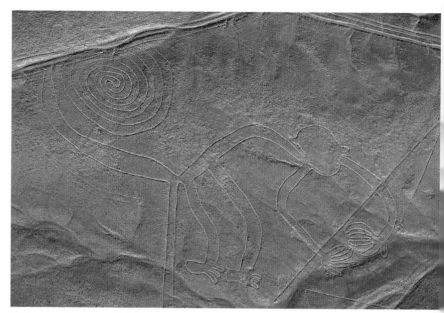

Truly an anthropomorphic figure, one can perhaps imagine that this enormous figure, as some believe, represents one of the "alien astronauts" that assisted in the construction of the more than 300 geometric figures on the ground. Since the true character of each one can only be realized from hundreds of metres in the air, and not on the ground, the 3,000-year-old figures present quite a paradox when ascertaining exactly how they were constructed without aerial support. (Wikimedia Commons/Diego Delso/delso.photo/License CC-BY-SA)

cherry-picked to promote a pre-conceived and perhaps even selfish enticing public attraction?

As the author[2] was quoted by David Malakoff, in a recent article on prehistoric timekeeping[5] in the western hemisphere, "There is a lot of poppycock out there, so you have to be careful," says P. Clay Sherrod, an astronomer with Arkansas Sky Observatories who has helped analyze celestial alignments at dozens of prehistoric mound sites in the Mississippi River Valley. "It's too easy for people to first come up with a conclusion" – that a site reflects an extinct culture's reverence for solar alignments, for example – "and then go out and cherry pick the data needed to support it."

5. The Archaeological Conservancy: *American Archaeology*, Celestial Timekeeping by D. Malakoff, **19**, #1, Spring 2015.

This sentiment was echoed by Anthony Aveni when he wrote, "… the problem at most North American sites is that you don't have much evidence, you only have alignments."

But among all the theories and amidst the data that might tell us otherwise, we must remember one hard fact about life in the ancient Americas: these durable and creative people "made their living" staying alive, chasing animals and subsequently planting and harvesting crops to sustain large villages of people. If necessity is indeed "the mother of invention", then the calendars and astronomical alignments were the tools by which making a living kept these civilizations afloat.

And therein we have both the temptations and the roadblocks to truly understanding the seemingly expansive array of sun worship sites, elaborate calendars and markers of the solstices, and our desire to think beyond the artificial context in which ancient people have been portrayed – savages with neither intuitive thinking nor technological motivation in their lives.

The astronomical implications of the American pre-Columbian cultures' symbolism, structures and sophistication, as seen through what we witness in the forests, jungles and deserts of the Western Hemisphere, may well speak to us otherwise.

Astrophysicist Cecilia Helena Payne and Professor H. N. Russell

David M. Harland

Born on 10 May 1900, Cecilia Helena Payne was the first of three children by Edward John Payne, an Oxford-educated London barrister, and Emma Leonora Helena Pertz, who had Prussian roots. Her parents had married late in life and settled in Wendover, Buckinghamshire. Her father was in his mid fifties when Cecilia was born, then died when she was four years old. The widow supported her family by working as a musician and selling artwork that she painted herself.

At age six, Cecilia started formal education at a small co-educational school. When 12, the family moved into London and she was enrolled at St. Mary's College, Paddington, which was strong on divinity. At 16, she was moved to the more modern St. Paul's Girls' School, which had laboratories for science.

Arthur Stanley Eddington. (Wikimedia Commons/George Grantham Bain Collection, Library of Congress, Washington, D.C.)

In 1919 she won a scholarship to Newnham, an all-female college at Cambridge. Her main subject was her childhood passion for botany, but she broadened her curriculum with physics and chemistry.

Earlier in the year, Arthur Stanley Eddington, who was Cambridge's leading astrophysicist, had mounted a scientific expedition to the island of Príncipe, off the west coast of Africa, to photograph the solar eclipse of 29 May. In his Theory of General Relativity of 1915, Albert Einstein introduced a new concept of gravitation in terms of how 'space-time' is warped by the presence of mass. He predicted that starlight travelling close to the Sun on its way to us would be deflected by a given angle; roughly twice the angle predicted by Newtonian gravity. The Príncipe expedition had taken photographs to assess this. At the end of that year Eddington presented the results, which supported the hypothesis, at a meeting in the Great Hall of Trinity College. Payne was in the audience and was so impressed that she joined the Newnham Astronomy Society and used its telescope to explore the heavens for herself.

At an 'open event' at the Observatory of Cambridge University some time later, Payne approached Eddington and told him of her desire to become an astronomer. When he encouraged her in this endeavour she added courses in astronomy to her curriculum.

In England, the best career she could hope for after university was a teaching position in a school for girls, so she set her sights on America where her chances of a career in astronomical research were better.

At the Centennial Meeting of the Royal Astronomical Society held in London in 1922, Payne attended a lecture by Harlow Shapley, a young and successful research astronomer who had recently become Director of the Harvard College Observatory. After the lecture, she approached him and said she would very much like to do postgraduate astronomy at his observatory. Shapley said she should write to him when the time came. Payne successfully completed her studies at Cambridge but was not given a degree owing to her gender. When she wrote to Shapley, he awarded her a Pickering Fellowship, a one-year award to allow women to pursue advanced studies at Harvard.

At that time, astrophysics was proving to be a worthy new form of astronomy by interpreting observations in terms of physical processes. With a firm grounding

Harlow Shapley. (National Academy of Sciences)

Cecilia Helena Payne. (Harvard College Observatory / Smithsonian Institution Archives)

in physics, chemistry and astronomy, and also an enthusiastic endorsement from none other than Eddington, Payne arrived at Harvard in the autumn of 1923 to carry out research under the supervision of Shapley. He suggested that she pick up the work of Henrietta Swan Leavitt, recently deceased, on variable stars,[1] but Payne had in mind a project of her own. She wanted to interpret stellar spectra in terms of recent theories of atomic structure. This was an entirely new kind of investigation for his observatory, but Shapley was enthusiastic.

Stellar Spectra

On the basis of 4,000 observations, in 1867 Pietro Angelo Secchi at the Observatory of the Roman College in Italy proposed a classification scheme for stellar spectra. Meanwhile, in Sweden the physicist Anders Jonas Ångström, who had spent a decade studying the Sun, published a chart in 1868 detailing 800 narrow dark lines in its spectrum. In 1880 in New York, Henry Draper obtained detailed photographs of the spectra of bright stars using a 28-inch reflector. On comparing the spectrum of the Sun to Draper's spectra, Edward Charles Pickering, Director of the Harvard College Observatory at that time, refined Secchi's classification scheme.

Henry Draper, c.1872. (Wikimedia Commons / William R. Howell (New York)

In 1885, Pickering arranged for an 'objective prism' to be installed at the front of a telescope in order to obtain a spectrum for each star in the field of view simultaneously, rather than photographing them one at a time. The resolution was low, but revealed vital characteristics. He then kicked off a vast campaign to compile a photographic catalogue of the positions and properties of stars. Dissatisfied with the performance of male assistants in analysing the objective prism spectra, he assigned the task to his Scottish housemaid, Williamina Fleming, in the belief that a woman would be more fastidious, which was true. She soon devised an entirely new classification scheme.

---

1. I discussed Henrietta Swan Leavitt's work at the Harvard College Observatory in *Yearbook of Astronomy 2021*.

Harvard College Observatory in 1899. The archive of glass photographic plates was housed in the appropriately named Brick Building. (Harvard College Observatory)

Since the brightest stars were over-exposed on the objective prism spectra, Pickering arranged for an 11-inch refractor to obtain spectra of such stars one at a time. In 1887 he hired Antonia Cætana Maury out of Vassar College, where she had scored well in physics, astronomy and philosophy, and put her to work on analysing the bright stars. With a wealth of very fine detail in their spectra, Maury was able to improve the scheme. The spectra from the objective prism were used to classify large numbers of stars in general terms and the detailed spectra of the bright stars yielded insight into how the classification scheme might be further refined.

Annie Jump Cannon graduated in 1884 from Wellesley College with interests in physics and astronomy, and then enrolled at the Society for the Collegiate Instruction of Women. Founded in 1879 as a Harvard Annex to enable professors to repeat their lectures to women, it became Radcliffe College in 1894. In 1896, aged 32, she joined Pickering's team. Stellar spectra had been arranged into a linear alphabetic sequence presumed to match how stars were formed and evolved, but as our understanding of such matters improved Cannon revised the ordering of the letters to OBAFGKM, with a subdivision number.[2] For example, the Sun is a G2 star.

---

2. As an *aide-mémoire*, when astronomers see this sequence they tend to mentally recite: 'Oh Be A Fine Girl Kiss Me'.

Some of the women who worked at Harvard College Observatory as 'computers'. On the far left is Henrietta Leavitt. In the foreground is Annie Jump Cannon, with Williamina Fleming standing behind her. On the far right is Antonia Maury. (Harvard College Observatory)

Draper's widow provided most of the funding for the photography and analysis, requesting only that the result be called the Henry Draper Catalogue. When finished in the early 20th century it encompassed the entire sky down to a photographic magnitude of about 9, and included spectroscopic classifications for over a quarter of a million stars. It was a tremendous resource for the new science of astrophysics. In 1910 the Draper classification was adopted as an international standard.

Atomic Structure

In 1913 Niels Bohr of the University of Copenhagen had applied the 'light quanta' conceived by Max Planck in 1900 to Ernest Rutherford's 1911 model of nuclear structure. In Bohr's model of the hydrogen atom a single negatively charged electron was bound to a positively charged nucleus comprising a single proton. There were distinct energy states, and the transitions between them matched the wavelengths of the lines observed in the hydrogen spectrum.

In 1921 Megh Nad Saha, a young Indian physicist in Calcutta (now Kolkata), applied thermodynamics to Bohr's atom and realised that the 'ionisation' states of atoms could yield information about the physical conditions in stars.[3] The higher the temperature, the greater the heat energy and the more readily the electrons in atoms were able to adopt excited states, thereby absorbing light to create the dark narrow lines at characteristic wavelengths called Fraunhofer lines If the temperature were sufficiently hot, electrons could escape altogether. Saha derived a mathematical relationship to predict the Fraunhofer lines which ought to be present at the very high temperatures in stars. In comparing his results against real spectra he found general agreement. This indicated that differences in the spectra of stars

Megh Nad Saha in 1921.
(Wikimedia Commons)

were indicative of differences in temperature rather than, as had been believed, differences in chemical composition. The spectral types signified a temperature scale, with the hottest stars being a blue-white hue and the coolest a reddish hue.

Payne's Research

Payne was familiar with Saha's work because Edward Arthur Milne at Cambridge had extended it to lower pressures. Before she set off for America, Milne had suggested she access the Henry Draper Catalogue and study the intensities of Fraunhofer lines in stellar spectra to obtain *quantitative* data for the ionisation of stars.

The research project that Payne wanted to pursue was to use Saha's relationship in reverse. Instead of using atomic theory to predict what lines would occur in the spectra of stars in terms of their temperatures, she would use stellar spectra to determine the physical conditions and chemical compositions of stars. This would enable her to determine the relative abundances of the elements in stars.

---

3. 'On a Physical Theory of Stellar Spectra', M. N. Saha. 1921, Proceedings of the Royal Society, **99**(A), 135–153.

Having selected certain Fraunhofer lines for analysis, she undertook the tedious task of estimating their intensities for hundreds of spectra. In the case of silicon, she identified four successive stages of ionisation (the neutral atom with 14 electrons, and then stripped of one, two and three electrons) and calculated the heat required to attain each of these states. This indicated that the temperatures of the O-type stars ranged from 23,000° to 28,000°. Element by element, she used their ionisation to calculate the temperatures of very hot stars classified as types O and B in the Harvard system.

Henry Norris Russell. (Wikimedia Commons/Princeton University/ The World's Work, 1921)

Impressed by Payne's work, Shapley urged her to aim for a doctoral rather than a master's degree. The problem was that Harvard had never awarded a Ph.D. to a woman. He convened a faculty committee, which drew up a written examination. On passing this on 10 June 1924, Payne became a Ph.D. candidate. When her Pickering Fellowship expired, the Rose Sidgwick Memorial Fellowship of the American Association of University Women funded her second year.

Pushing on, Payne applied her knowledge of statistical mechanics to estimate the number of atoms necessary to make each absorption line, and thereby determine the abundances of the 18 elements that are most common in the atmospheres of various types of stars. Her results were in remarkable agreement with expectations, but there was also a major surprise.

The prevailing view was that stellar atmospheres were composed of vapours of heavy elements. In 1914 Henry Norris Russell at Princeton University had compared the Fraunhofer lines in the Sun with the most common elements present in Earth's crust, particularly oxygen, silicon, aluminium and iron, and found a match for relative abundances.[4] This result was consistent with both the theory that the Sun and its planets formed in the same nebula and also the rival theory where the planets condensed from material torn from the Sun during a near collision

---

4. 'The Solar Spectrum and the Earth's Crust', H. N. Russell. 1914, *Science*, **39**, 791–794.

with another star. On the presumption that the Sun is a normal star, all stars could reasonably be expected to have the same relative abundances.

Although Payne found the terrestrial elements to be present in stars, the overall composition was dominated by the two elements having the lightest atomic numbers: hydrogen and helium. Indeed, the amount of hydrogen in stars seemed to be a million times greater than predicted. Remarkably, her analysis showed that all elements heavier than helium were present in merely *trace* amounts in stars.

In December 1924, Shapley sent a draft of Payne's analysis of stellar abundances to Russell, who had supervised his own Ph.D. and was the acknowledged American expert on the nature of stars. In replying on 14 January 1925, Russell praised Payne's work but warned it was "clearly impossible that hydrogen should be a million times more abundant". In view of this criticism, Payne tempered her conclusions when writing a paper in February for the *Proceedings of the National Academy of Sciences*, pointing out that the "improbably high" abundances of hydrogen and helium were "almost certainly not real".[5]

In speaking of the temperature of a star, Payne was referring to the temperature of its visible 'surface', which is really only a distinct layer in its 'atmosphere'. In the case of the Sun, the visible surface is known as the 'photosphere' because it is the source of light. Immediately above it is the 'chromosphere', which is the layer of cooler gas which imprints the absorption lines on the 'white light' from the photosphere. For this reason, the chromosphere is also referred to as the 'reversing layer'.

At each step in her work, Payne had written up her methodology, results, and reflections. As a result, she required only 6 weeks to write her doctoral thesis, entitled 'Stellar Atmospheres: A Contribution to the Observational Study of High Temperature in the Reversing Layers of Stars'. She outlined the procedures she had devised for analysing stellar spectra, gave the temperature scale she had calibrated, and summarised the inferred abundances of the elements in the stars. And just as in her paper, she questioned the validity of the hydrogen and helium results.[6]

Four years after he criticised Payne's claim for hydrogen in stars, in 1929 Russell realised that she was correct.[7] With improved data for the Sun (and very likely her

---

5. 'Astrophysical Data Bearing on the Relative Abundances of the Elements', C. H. Payne. 1925, *Proceedings of the National Academy of Sciences*, **11**, 192–198.

6. Strictly speaking, Payne gained her doctorate from Radcliffe College, the annex of Harvard University for women.

7. 'On the Composition of the Sun's Atmosphere', H. N. Russell. 1929, *Astrophysical Journal*, **70**, 11–82.

thesis in the back of his mind) he realised that many difficulties in interpreting the solar spectrum could be overcome if "the solar atmosphere really does consist mostly of hydrogen" with there being "hardly more than a smell of metallic vapours in it." When comparing the relatively cool Sun to the far hotter OB stars that Payne had investigated, Russell reported "a very gratifying agreement".

Unfortunately for Payne, Russell's stature in the field ensured that it was *his paper* which became the standard reference for hydrogen being the principal constituent of stars.

Nevertheless, Otto Struve, renowned astronomer of the University of Chicago, later wrote that Payne's was "undoubtedly the most brilliant Ph.D. thesis ever written in astronomy. It was published at the right time and contained the right kind of information to stimulate many new investigations".[8]

The surprise was that it is possible for an element to be common in a star, yet be hard to detect. An atom that is fully ionised will possess no electrons to jump from low energy states to higher energy states, in the process absorbing energy and contributing to Fraunhofer lines. Payne's calculations showed that for the Sun the vast majority of the hydrogen must be ionised. Heavier atoms have so many Fraunhofer lines in the solar spectrum because it is almost impossible to strip away all of their electrons. The worst case is iron, whose neutral atom has 26 electrons. That was why astronomers thought stars were rich in iron. The ionised hydrogen and helium were not showing up. But Payne's analysis had inferred their presence. It was the abundances of hydrogen and helium that made her thesis more than a

HARVARD OBSERVATORY MONOGRAPHS
HARLOW SHAPLEY, Editor

No. 1

STELLAR ATMOSPHERES

A CONTRIBUTION TO THE OBSERVATIONAL
STUDY OF HIGH TEMPERATURE IN THE
REVERSING LAYERS OF STARS

BY

CECILIA H. PAYNE

PUBLISHED BY THE OBSERVATORY
CAMBRIDGE, MASSACHUSETTS
1925

Title page of *Stellar Atmospheres* by Cecilia Payne, the first person to earn a Ph.D. in astronomy from the Radcliffe College of Harvard University. (Harvard Observatory Monographs, no. 1, 1925; Harvard College Observatory/Radcliffe College/Open Library)

8. *Astronomy of the Twentieth Century*, Otto Struve and Velta Zebergs. 1962, Macmillan, 220–221.

straightforward verification of Saha's ionisation theory. We now know hydrogen and helium account for 98% of the mass of the Sun.

Normally, the publications of the Harvard College Observatory were distributed only to subscribing institutions, but to mark his first female student getting a doctorate Shapley started a *Monograph* series, leading with Payne's 215-page thesis. In the realm of highly technical reports it was a best seller. Indeed, by shipping all 600 copies in 3 years at a price of $2.50 each it returned a profit.

Ironically, when the significance of Payne's work was recognised she was listed in the renowned biographical encyclopaedia *American Men of Science*.

Postscript

Following her graduation, and now a research fellow at Harvard, Payne studied high luminosity stars to investigate the supposed spiral structure of the galaxy. In 1927 she was appointed a technical assistant to Shapley, enabling her to give lectures.

Payne became an American citizen in 1931. Two years later, on a tour of astronomical establishments in Europe, she attended a conference in Göttingen, Germany, and encountered Sergei I. Gaposchkin, a stateless Russian astronomer. On returning to America, she helped him to obtain a visa to enter the United States. They married in March 1934 and had three children. In 1938 Shapley invented a staff position for her with the title of 'Astronomer'.

Then in 1956, with Donald H. Menzel having succeeded Shapley as director, Payne was the first woman to be promoted to a professorship from within Harvard's Faculty of Arts and Sciences. If she had been a man, her advancement would certainly have been more rapid. Later, as Chair of the Department of Astronomy, she became the first woman to head a department at that university.

Payne spent her entire career at Harvard and remained scientifically active. She wrote a number of highly regarded volumes, including: *The Stars of High Luminosity* in 1930, *Variable Stars* in 1938, *Variable Stars and Galactic Structure* in 1954, *Stars in the Making* in 1952, *Introduction to Astronomy* in 1954, *The Galactic Novae* in 1957, and *Stars and Clusters* in 1979. On retiring from active teaching in 1966 as an Emeritus Professor, she joined the staff at the Smithsonian Astrophysical Observatory.

In 1934 Payne became the first recipient of the Annie Jump Cannon Award in Astronomy. In 1961 she received the Rittenhouse Medal from the Rittenhouse Astronomical Society at the Franklin Institute. Her *alma mater*, Radcliffe College, gave her its Award of Merit in 1952. And the American Astronomical Society awarded her its H. N. Russell Prize in 1976 in recognition of her lifetime of excellence in astronomical research.

Payne died at home on 7 December 1979, aged 79. Distraught at the loss of his wife, Gaposchkin suffered a rapid decline in health, survived a stroke, and died on 17 October 1984.

Shortly before her death, she had her autobiography privately issued as *The Dyer's Hand*. It was reprinted by Cambridge University Press in 1984 with the title *Cecilia Payne-Gaposchkin: An Autobiography and Other Recollections*.

Further Reading

The Glass Universe: The Hidden History of the Women Who Took the Measure of the Stars by Dava Sobel, 4th Estate, 2016.

What Stars Are Made Of: The Life of Cecilia Payne-Gaposchkin by Donovan Moore, Harvard University Press, 2020.

An Autobiography and Other Recollections by Cecilia Payne-Gaposchkin, Cambridge University Press, 1984.

The Astronomers' Stars
The Terrible Twos

Lynne Marie Stockman

A binary star is actually two stars, orbiting around their common centre of gravity. The term 'binary' was first used in this context by astronomer William Herschel in the nineteenth century (Herschel 1802):

> If a certain star should be situated at any, perhaps immense, distance behind another, and but very little deviating from the line in which we see the first, we should then have the appearance of a double star. But these stars, being totally unconnected, would not form a binary system. If, on the contrary, two stars should really be situated very near each other, and at the same time so far insulated as not to be materially affected by the attractions of neighbouring stars, they will then compose a separate system, and remain united by the bond of their own mutual gravitation towards each other.

A visual binary is detected directly as two stars close together in the sky, but many binary star systems cannot be resolved into separate stars. Spectroscopic binaries are inferred indirectly through careful analysis of their spectra. Astrometric binaries are detected from the periodic 'wobble' in the position of the brighter star caused by the gravitational pull of a fainter companion. Eclipsing binaries orbit each other in a plane along our line of sight, leading to fluctuating brightnesses as one moves in front of the other. The masses of binary components can be calculated, which allows scientists to derive other stellar characteristics, like radius and luminosity. All of these quantities are necessary to model stellar evolution. Binaries are the astrophysicist's playground.

Born at his family's farm in Ontario, Canada, **John Stanley Plaskett** (1865–1941) had an interrupted early education as he was often needed to work at home. He was interested in all things mechanical and learned the machinist's trade, working in Canada and the United States before being hired by the physics department of the University of Toronto. Here he constructed specialised equipment both for the laboratory and the lecture theatre. This exposure to higher education kindled a fascination for physics and he began his bachelor's degree studies at the age of 30,

The dome and mount for the great 72-inch telescope of the Dominion Astrophysical Observatory were completed in 1916 but the observatory was not operational until 1918 due to manufacturing issues with the mirror. The telescope was the second-largest reflector in the world at the time it was finished and is now known as Plaskett's Telescope. The DAO has been open for public viewing almost continuously since June 1918. This image of the newly-constructed dome dates to around 1916. (Dominion Astrophysical Observatory, Victoria, BC; provided by Friends of the Dominion Astrophysical Observatory, **centreoftheuniverse.org**)

graduating with honours in mathematics and physics in 1899. He stayed at the university until 1903 when he was appointed to the staff at the new Dominion Observatory in Ottawa. Here he not only carried out astronomical research but used his mechanical skills to build and improve the equipment. The Dominion Astrophysical Observatory (DAO) in Victoria was opened in 1918 under Plaskett's direction and he remained there until 1935. His research was often concerned with spectroscopy and received many accolades, including the Gold Medal of the Royal Astronomical Society in 1930 (Anonymous 1942).

Still known as **Plaskett's Star**, the sixth-magnitude variable V640 Monocerotis (BD+06 1309) was studied by the astronomer in 1921 and 1922 with spectroscopy revealing its true binary nature (Plaskett 1922):

> The mass of 6° 1309 is then about 4 times that of any previously known stellar system and with the very high temperature and surface luminosity of the Oe and Oe5 stars must be exceedingly bright. It will be of interest if we cannot obtain some estimate of its absolute magnitude and parallax.

Plaskett had discovered a monster. Values from the *Gaia* mission give an annual parallax of 0.6583 milli-arcseconds which implies a distance of approximately 1520 pc. Analysis of *CoRoT* data revises the spectral types to O8I (bright supergiant) and O7.5III (giant) which suggests that both stars have luminosities of tens to hundreds of thousands times that of the Sun. Orbiting each other in a period of just over 14 days, the two stars making up Plaskett's Star still comprise one of the most massive binaries yet discovered. Research continues to reveal new secrets about this unusual object.

Like his older colleague Plaskett, **Joseph Algernon Pearce** (1893–1988) was born in Ontario, Canada. His university studies at the University of Toronto were interrupted by the First World War when Pearce enlisted in the 58th Infantry Battalion, serving in France and Belgium before being injured in 1916. He achieved the rank of major by the end of the war and returned to Toronto to complete his undergraduate degree in mathematics and physics. After finishing his master's degree, he was awarded a fellowship to Lick Observatory where he was able to work toward his doctorate, earning that in 1930. Much of his time at Lick was dedicated to calculating orbits of comets and asteroids, and minor planet 3304 is named in his honour. In 1924, Plaskett, then director of the DAO, invited Pearce to join the staff in Victoria where the two astronomers embarked on a project to determine the structure of the galaxy. They established that the centre of the galaxy was in

In this artist's impression, Pearce's Star (AO Cassiopeiae) appears over the limb of a hypothetical circumbinary planet. The primary star is larger but less massive than the secondary star; it has filled its Roche lobe and is transferring material to its companion. (Lynne Marie Stockman)

the direction of Sagittarius and that stars nearer the galactic centre moved faster than those further out, thus confirming earlier theories of galactic differential rotation. The DAO was a world leader in the study of massive binaries and Pearce became one of the first astronomers to calculate accurate masses and radii of the components of these systems. Pearce was promoted to assistant director of the DAO following Plaskett's retirement and eventually director in 1940 where he remained until he resigned in favour of another colleague in 1951. He stayed on at the observatory as Director Emeritus, retiring seven years later. He was a member of many learned societies, including the International Astronomical Union, where he served on several commissions relating to stellar spectra, photometric double stars and radial velocities. He died aged 95 (Wright 1989).

Also known by the variable star designation AO Cassiopeiae, **Pearce's Star** was revealed to be a binary system in spectrograms taken at Mount Wilson Observatory between 1914 and 1916. Pearce observed the star ten years later, in 1924 and 1925. Combining his results with the earlier Mount Wilson data, Pearce was able to determine the dimensions of the binary system (Pearce 1926):

The subject of this paper is an eclipsing variable of exceptional mass and exceedingly high temperature. With actual masses of 36.3 and 33.8 times the sun, *H.D. 1337*, spectral type O8.5n, stands at present at the head of the limited list of fourteen systems whose absolute dimensions have been derived by combining the elements their spectroscopic and photometric orbits.

Although later observations and improved modelling have suggested smaller masses and radii for the two objects, they are still large, hot, O-type stars – one main sequence and the other a giant or bright giant. They revolve around each other in just 3.52 days, a value first calculated by Pearce. Because the two stars are so close to each other, they are markedly ellipsoidal in shape. Pearce's Star is thought to be either a contact binary (where the two stars actually touch each other) or a semi-detached binary (where one of the two stars has filled its Roche lobe).

Centaurus X-3 was discovered in 1967 from data gathered by sounding rockets. Later observations were performed with *Uhuru*, the first dedicated X-ray detection satellite, and it was found that Centaurus X-3 was actually an X-ray variable star with a 4.87-second period. However, the period varied slightly over time and astronomers could not immediately explain why (Giacconi 1971). Whilst working at the European Southern Observatory at La Silla, Chile, Polish astronomer **Wojciech Krzemiński** (1933–2017) used a photometer to examine the region of space immediately adjacent to the X-ray source and found the optical companion that had been predicted two years earlier (Krzemiński 1974):

> The optical component of the pulsating binary X-ray source Cen X-3 is identified with a faint, <V> = 13.35, heavily reddened variable star lying slightly outside the error box of the 3U catalog.

The optical component is now called *Krzemiński's Star*. This star is an O-type giant with a mass of just over 20 solar masses and a radius of approximately 12 solar radii. Centaurus X-3 is comprised of Krzemiński's Star and a neutron star, the first X-ray pulsar to be discovered. The orbital period of the two objects is just over two days. The X-rays are thought to be generated by material from Krzemiński's Star overflowing its Roche lobe onto an accretion disk which eventually spirals down onto the neutron star (Naik et al 2011).

Physics postgraduate student **Russell Alan Hulse** and his thesis adviser astronomer **Joseph Hooton Taylor** undertook a high-sensitivity pulsar survey at Arecibo Observatory in the early 1970s. They discovered a number of new pulsars with one standing out in particular (Hulse and Taylor 1975):

We wish to report the detection of an unusual pulsar discovered during the course of a systematic survey for new pulsars being carried out…at the Arecibo Observatory in Puerto Rico. The object has a pulsation period of about 59 ms—shorter than that of any other known pulsar except the one in the Crab Nebula—and periodic changes in the observed pulsation rate indicate that the pulsar is a member of a binary system with an eccentric

Arecibo Observatory was a radio telescope with a 305-metre diameter dish constructed within a naturally-occurring sinkhole on the island of Puerto Rico. (The automobile at the bottom right gives a sense of scale.) Opened in 1963, the observatory was used initially by the United States Department of Defense to study Earth's ionosphere. The DoD transferred the facility in 1969 to the National Science Foundation (NSF). Arecibo was used for both radar and radio astronomy, and played a part in the Search for Extraterrestrial Intelligence (SETI) project. Despite the observatory contributing to numerous important scientific discoveries throughout the latter part of the twentieth century and beginning of the twenty-first, funding for the observatory waned and the dish collapsed in December 2020 before it could be safely decommissioned. (NAIC Arecibo Observatory, a facility of the NSF)

orbit of 0ᵈ.3230 period. Thus for the first time it is possible to observe the gravitational interactions of a pulsar and another massive object....

In analysing the timing data, they concluded that the object, catalogued as PSR 1913+16, was a binary, with one component a pulsar and the other a compact object such as a neutron star or a black hole. A later determination of the mass confirmed the invisible component as a neutron star.

The *Hulse-Taylor Pulsar*, as the object came to be called, was a sensation. By the 1980s, Taylor and his colleague Joel Mark Weisberg were using the first known binary pulsar as a gravitational laboratory to test Albert Einstein's theory of general relativity. Years of pulse timing data were used to specify the orbital parameters and masses of the two objects, to determine the decay rate of the orbit, to rule out several other competing theories of gravitation, and to test solutions of the relativistic two-body problem (Taylor and Weisberg 1989):

The orbit is losing energy within 1% of the rate predicted for gravitation radiation damping in general relativity—which we interpret as incontrovertible evidence for the existence of gravitational waves.

Gravitational waves were finally observed directly in 2015 when the Laser Interferometer Gravitational-Wave Observatory (LIGO) in the United States detected the aftermath of two black holes merging.

The Hulse-Taylor Pulsar had a final gift to bestow upon its discoverers. In 1993, Hulse and Taylor were jointly awarded the 1993 Nobel Prize in Physics for their discovery of this new type of pulsar, with the prize committee citing the pulsar's extraordinary potential in the study of gravitation (Press Release 1993).

In 1977, American astronomers **Charles Bruce Stephenson** (1929–2001) and **Nicholas Sanduleak** (1933–1990) published a list of hydrogen alpha emission stars in the Milky Way. Number 433 on this list was simply described as magnitude +13.5 and having sharp Hα emission lines (Stephenson and Sanduleak 1977). Other observers soon noticed that it was variable in many different wavelengths, from radio to X-ray, and that details of its spectrum seemed to change, with large Doppler shifts apparently suggesting that the object was moving to and fro at tens of thousands of kilometres per second. The *Stephenson-Sanduleak Object* or SS 433 was becoming notorious for its unusual behaviour. Theories as to the nature of the object came thick and fast. Veiled pulsar? Jets? Some kind of new stellar remnant? A binary? Its spectrum defied easy explanation and SS 433 was described as the most exotic star system yet discovered.

Hundreds of research papers have been published relating to SS 433, detailing observations and offering theories as to what it actually is. Current thinking is that SS 433 is an eclipsing X-ray binary system, consisting of a compact object, probably a black hole, and an A-type mass donor star, so-called because it is 'donating' its mass to an accretion disk which then spirals down onto its companion. The compact object also possesses collimated relativistic jets which are outflowing at speeds of 25% the speed of light. The binary has a 13-day period and the accretion disk precesses over the course of 162 days (Hillwig et al 2004). It is the prototype microquasar, a smaller, less powerful version of a quasar. And although astronomers have given it a label, it remains one of the most mysterious objects in the universe.

On 15 June 1982, the Central Bureau for Astronomical Telegrams announced the discovery of 'Twin White Dwarfs' (IAUC 3703):

N. Sanduleak and P. Pesch, Warner and Swasey Observatory, report the discovery on a low-dispersion objective-prism plate of an astrophysically important binary system containing nearly identical probably DA white

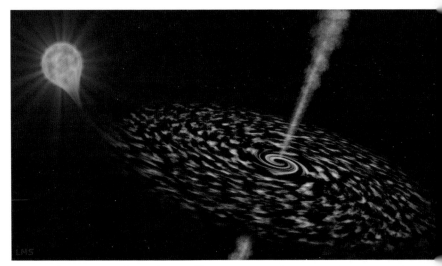

This artist's impression shows the variable X-ray source SS 433 as an A-type supergiant star circling a black hole. Gas streams from the donor star onto an accretion disk surrounding the degenerate object, with relativistic jets of hydrogen racing away along the black hole's axis of rotation at over 25% the speed of light. (Lynne Marie Stockman)

dwarfs…The stars can be readily identified on the Sky Survey prints, where they lie 6.0 mm east-southeast of the star SAO 46477.

At the time, Sanduleak and **Peter Pesch** were collaborating on a decade-long project entitled 'The Case Low-Dispersion Northern Sky Survey' of interesting faint spectra of stars and galaxies at high galactic latitudes when they came across the object (Stephenson 1991). Although not the first binary white dwarf system discovered, the **Sanduleak-Pesch Binary** was the brightest and hottest to date and thus, the youngest.

Other researchers immediately followed up on the unusual object. It was quickly determined that the stars were young Population I objects and aged approximately 500 million years. One dwarf (component A) was hotter and more massive than the other (component B) and it was calculated that the progenitor stars were 8 and 4 solar masses respectively (Greenstein et al 1983). But the cooler B component still had a little surprise (Maxted 2000):

WD 1704+481 is a visual binary in which both components are white dwarfs. We present spectra of the Hα line of both stars which show that one component (WD 1704+481.2 = Sanduleak B = GR 577) is a close binary with two white dwarf components. Thus, WD 1704+481 is the first known triple degenerate star.

Component B gives two white dwarfs for the price of one! Whilst component A seems to be a typical white dwarf, components B and C make up a close binary (less than an astronomical unit apart) with a period of only 3.5 hours. Careful examination of the data suggests that component B, which is less massive than C, is composed of helium whilst C is made up of carbon and oxygen. Perhaps the Sanduleak-Pesch Binary should be renamed the **Sanduleak-Pesch Trinary**?

Acknowledgements

This research has made use of NASA's *Astrophysics Data System Bibliographic Services*, operated at the Harvard-Smithsonian Center for Astrophysics, Cambridge, Massachusetts, USA, and the *SIMBAD* astronomical database, operated at CDS, University of Strasbourg, France. The author would like to thank Dr Ben Dorman from the Friends of the Dominion Astrophysical Observatory for providing a selection of wonderful images of the DAO and Dr David Harper for his enthusiastic encouragement and helpful comments.

References

[Anonymous] 1942, 'Obituary: John Stanley Plaskett, C.B.E.', *Monthly Notices of the Royal Astronomical Society*, **102** (2), 70–73.

Giacconi, R., Gursky, H., Kellogg, E., and 2 others. 1971, 'Discovery of Periodic X-ray Pulsations in Centaurus X-3 from UHURU', *The Astrophysical Journal*, **167** (2), L67–L73. **doi.org/10.1086/180762**

Greenstein, J.L., Dolez, N., Vauclair, G. 1983, 'Physical properties and evolution of the two white dwarfs in the Sanduleak-Pesch binary', *Astronomy & Astrophysics*, **127**, 25–28.

Herschel, William. 1802, 'XVIII. Catalogue of 500 new Nebulæ, nebulous Stars, planetary Nebulæ, and Clusters of Stars; with Remarks on the Construction of the Heavens', *Philosophical Transactions of the Royal Society of London*, **92**, 480–481.

Hillwig, T.C., Gies, D.R., Huang, W. and 4 others. 2004, 'Identification of the Mass Donor Star's Spectrum in SS 433', *The Astrophysical Journal*, **615** (1), 422–431. **doi.org/10.1086/423927**

Hulse, R.A., Taylor, J.H. 1975, 'Discovery of a Pulsar in a Binary System', *The Astrophysical Journal*, **195** (2), L51–L53. **doi.org/10.1086/181708**

IAU Circular No. 3703 (1982).

Krzemiński, W. 1974, 'The Identification and UBV Photometry of the Visible Component of the Centaurus X-3 Binary System', *The Astrophysical Journal*, **192** (3), L135–L138. **doi.org/10.1086/181609**

Maxted, P.F.L., Marsh, T.R., Moran, C.K.J., and 1 other. 2000, 'The triple degenerate star WD 1704+481', *Monthly Notices of the Royal Astronomical Society*, **314** (2), 334–337. **doi.org/10.1046/j.1365-8711.2000.03343.x**

Naik, Sachindra, Paul, Biswajit, Ali, Zulfikar. 2011, 'X-ray Spectroscopy of the High-Mass X-ray Binary Pulsar Centaurus X-3 Over its Binary Orbit', *The Astrophysical Journal*, **737** (2), 79–86. **doi.org/10.1088/0004-637X/737/2/79**

Pearce, J.A. 1926, 'The Absolute Dimensions of the O-Type Eclipsing Variable H.D. 1337', *Publications of the Dominion Astrophysical Observatory Victoria*, **3** (13), 275–306.

Plaskett, J.S. 1922, 'The Spectroscopic Orbit of B.D. 6° 1309', *Publications of the Dominion Astrophysical Observatory Victoria*, **2** (4), 147–158.

[Press Release] 1993, 'The Nobel Prize in Physics 1993', The Royal Swedish Academy of Sciences.

Stephenson, C.B., Sanduleak, N. 1977, 'New H-alpha Emission Stars in the Milky Way', *The Astrophysical Journal Supplement Series*, **33** (4), 459–469. **doi.org/10.1086/190437**

Stephenson, C. Bruce. 1991, 'Nicholas Sanduleak, 1933–1990', *Bulletin of the American Astronomical Society*, **23** (4), 1491–1492.

Taylor, J.H., Weisberg, J.M. 1989, 'Further Experimental Tests of Relativistic Gravity using the Binary Pulsar PS 1913+16', *The Astrophysical Journal*, **345** (1), 434–450. **doi.org/10.1086/167917**

Wright, Kenneth O. 1989, 'Joseph A. Pearce, 1893–1988', *Journal of the Royal Astronomical Society of Canada*, **83** (1), 3–7.

Mission to Mars
Countdown to Building a Brave New World
It's Life, But Not as We Know It

Martin Braddock

Introduction

The last article entitled 'The Right Stuff at the Right Time' in *Yearbook of Astronomy 2024* introduced the concept of the role of humanity in an increasingly technologically advancing world where designer humans may become a reality, if not a prerequisite for adaptation to extreme environments. The feasibility of Martian colonisation by the end of the 2020s is now perhaps unrealistic, though the ambition remains to land humans on Mars by the early 2030s.[1] In this article, I will introduce the concept of humanoid robots as our partners for Mars colonisation and habitation. The development and deployment of robotics to assist humans working in, or at distance from extreme environments, such as the nuclear industry, high and low temperature environments, in offshore maintenance, the mega-construction industry and deep mining has rapidly progressed and may be considered standard practice in many projects (Takahashi et al 2021, Sayed et al 2022). Robots are required to reduce or eliminate the risks associated with human presence, usually by reducing or removing the requirement for people to enter hazardous environments thus reducing or removing exposure to the hazard. In many manufacturing industries and often in confined spaces, robot-generated productivity far exceeds that which could be provided by humans, for example the food and automotive manufacturing industries, the tourist and on-line supply chain industry.

Robots in Space

Clearly the extent to which human beings may extend and persist into space is determined by the frailty of the human body, both physically and psychologically. Mars exploration in the early 2020s has focussed on the deployment of the rovers *Perseverance* and *Zhurong* from NASA's *Mars 2020* and China's National Space

---

1. Mars Colonization Timeline, **humanmars.net/p/mars-colonization-timeline.html**

Computational science reaching out. (Pixabay)

Agency's *Tianwen-1* missions respectively. Given the enormous challenges in getting to and landing on Mars, for technical and human safety requirements, the initial phases of space colonisation will be likely fully automated[2] (Campa et al 2019), with the assembly of pre-fabricated habitats including bio-regenerative life support systems conducted by robots utilising additive manufacturing methodology with minimal human intervention (Bier et al 2021, Flinders et al 2022).

Negating the extreme environment in such an off-world situation will be guided by robotics and terrestrial exploration of caves with unmanned aerial vehicles (Petracek et al 2021) and future capability for robots to climb on asteroids (Ribeiro et al 2023) are two examples paving the way for fully autonomous exploration. The recent discovery of water-ice and lava tubes on the Moon and Mars along with the development of *in situ* resource utilization technology has focused rover-based surface missions towards terrain exploration with the goal of siting a habitat location. To date, construction of three-dimensional terrain maps derived from a planetary or lunar orbiter has insufficient resolution and the recent development of visual simultaneous localization and robotic mapping employing a stereo camera system

2. Britt, H (2020), 'Construction Robots Will Have to Colonize Mars Before Humans', thomasnet.com/insights/construction-robots-will-have-to-colonize-mars-before-humans

An AI generated woman.
(Pixabay)

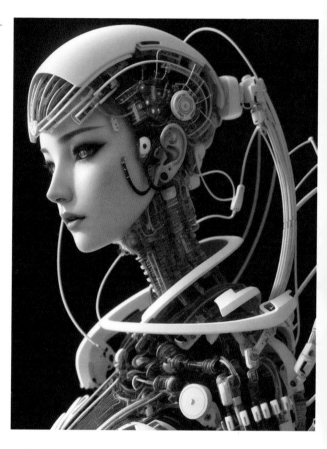

on a rover offers a potential solution (Hong et al 2021). A recent study proposes to link rovers, lake landers and even submersible vehicles through a mesh topology network exploring lava tubes and caves on Mars where machines work together as a team independently from human input (Fink et al 2023) providing solutions.

Advances in terrestrial biotechnology, electrical engineering and robotics science, machine-based learning artificial intelligence (MBL-AI), and the likely requirement for MBL-AI in the automated construction of extra-terrestrial habitats, lead to a future-world potential of human-robot hybrids (humanoids) as deep space voyagers. Despite a recent report on the development of sentient AI,[3] the

3. De Cosmo, L (2022), 'Google Engineer Claims AI Chatbot Is Sentient: Why That Matters', **scientificamerican.com/article/google-engineer-claims-ai-chatbot-is-sentient-why-that-matters**

development of machines with general artificial intelligence is believed to be many years away. However, simple responses in chatbots, and services such as Amazon's *Alexa* are in mass use today and our ability to generate life-like representations of human beings has existed for a number of years.[4] The creation of the humanoid robot Sophia in 2016,[5] subsequently given citizenship of Saudi Arabia in 2017, demonstrates the principle, and as of early 2023, there are four examples of robots which closely resemble humans.[6]

Fast forward into a future where human beings are chipped with radio-frequency IDentification (RFID) transponders, made available originally in 2005 to monitor cardiac function and now able to monitor total human vital signs. Space explorers, lunar and Martian colonists have successfully utilised smart clothing to sense their environment. A new science of bio-robotics has emerged where the integration of biology, biotechnology and robotics enables robots to feel, smell, hear and see in order to react to changing environments. When partnered with human beings, bio robots will be able to act as sentinels alerting humans physically or digitally via RFIDs to take evasive action, for example, to predict sudden increases in temperature or release of noxious substances. Some examples shown in Table 1 include the development of an electronic skin (E-skin) comprising sensing modules, which can be attached to a variety of surfaces. In Australia, researchers have developed an artificial skin that responds to pain mimicking the body's ability to provide immediate feedback when pressure, heat, or cold reach a certain threshold (Rahman et al 2020). Towards a future where robots are provided with warning signals, researchers at the Chinese University of Hong Kong have created an artificial skin which can change colour to simulate bruising (Qui et al 2021), using spiropyran, a molecule that changes colour from pale yellow to bluish-purple when exposed to mechanical stress. Scientists in France have developed an artificial nose and incorporated the device into a robot that can detect survivors in rescue operations and teams in Israel have created an optical nose that uses MBL-AI to detect odour signatures and can distinguish the aromas of red wine, beer, and vodka (Shumeiko et al 2021) and use the desert locust's primary olfactory apparatus to detect pure and mixtures of odorants (Neta et al 2023). Lastly, how do

---

4. Nelson, B (2023) 25 Real-Life Robots That Already Exist, **rd.com/article/real-robots**

5. 'Meet Sophia, the robot that looks almost human', *National Geographic*, #May 2018, **nationalgeographic.com/photography/article/sophia-robot-artificial-intelligence-science**

6. Orlando, A (2023), '4 Robots That Look Like Humans', **discovermagazine.com/technology/4-robots-that-look-like-humans**

robots 'see' their environment? Using a combination of light detection and ranging (LiDAR) and pan-tilt-zoom (PTZ) cameras, robots are able to orientate themselves in any environment, even to the point of over-coming obstacles.

Simple Features of Human Robots

Parameter	Example	Comments
Identity	Robot Sophia created	Given Saudi Arabian citizenship in 2017
Monitoring	Cardiac function with RFIDs	Senses and sentinels for avoiding hazards and extreme environments
Touch	E-skin sensing pain, bruising colour change	
Smell	Artificial nose	
Sight	LIDAR, PTZ	

(M. Braddock)

Life, but not as we know it?

The examples of robots which have been made to closely resemble human beings raise several interesting issues. First, do robots have rights, similar to animals and human beings (Gordon and Pasvenskiene 2021)? To answer this question from an anthropic perspective of being human, one may ask do machines have consciousness and as a consequence of consciousness, feeling and emotions? The answer, is no and lies in the anthropic viewpoint that though machines can sense the environment and react accordingly, robots may be able to express go (i.e. happiness) or stop (i.e. sadness) motions, but not feel the emotions in the way that human beings do. Intriguingly an online survey of 439 lay people suggests that robots should have rights, for example access to power and the ability to be updated rather than own property or have the ability to vote, suggesting that this debate will benefit by a common understanding of the future potential of robots to take on more and more human activity (De Graaf et al 2021). The second issue asks how comfortable we are as human beings to be guided by humanoid robots that look like and may be indistinguishable from ourselves in the future? In the 1970s, Japanese roboticist Masahiro Mori coined the term 'bukimi no tani genshō' which translates into 'uncanny valley' (Mori 2012), and describes a concept that as robots appear more human-like, they acquire a persona which makes them appealing to humans but only up to a limit and this effect may vary with the age of the person interacting with the robot with older adults showing lesser concern.

At the boundary of entering the uncanny valley, human acceptance decreases and humans develop a sense of unease and mistrust. This is illustrated in the figure

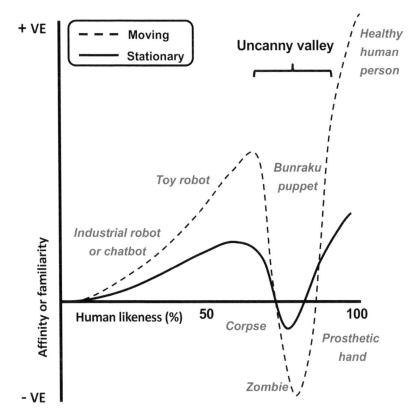

Representation of the uncanny valley and human emotive response to robots and their human-like appearance. (M. Braddock)

which shows hypothetical emotional responses against anthropomorphism of either a moving, animated object or a stationary, inanimate object. In 2021 the requirement for maintenance of personal space of both humans and human-like avatars has been modelled (Tootell et al 2021) and reported that discomfort-by-distance functions for both humans and avatars were closely aligned; in other words don't get too close to me! This reinforces the notion that humans need personal space, however, in contrast humans are a social species and robots as a countermeasure to loneliness have been studied (Jung and Hahn 2023). The need for personal space may not just be from other humans, but may also reflect the need for space from human-like objects and this could become important for the future design of humanoid robots both on Earth and in space for sustaining an effective partnership, especially in confined spaces.

The science of biorobotics is here today. Whether we like it or not, find it a threat to our livelihoods or an invaluable assist with our daily lives, robots present a potential solution to exploring inhospitable environments on Earth and in space. More importantly perhaps, biorobotics may teach us more about ourselves and what it means to be human.

As quoted by former astronaut John Glenn[7]

'I think sometime we will go to Mars and I think we'll explore it with humans sometime, but I think it's really wise to do all the robotic exploration ahead of time and learn as much as possible. Once we have learned as much as possible with the robots, then that's the time to send people, and let them then continue the research that the robots have started'.

References

Bier, H., Vermeer, E., Hidding A., Jani, K. 2021, 'Design-to-Robotic-Production of Underground Habitats on Mars', *SPOOL*, **8**, #2, 31–38.

Campa, R., Szocik, K., Braddock, M., 'Why space colonization will be fully automated', *Technological Forecasting and Social Change*, **143**, (June 2019), 162–171.

De Graaf, M. M. A., Hindriks, F. A. and Hindriks, K. V., 'Who Wants to Grant Robots Rights?', *Frontiers in Robotics and AI*, **8**, 2022, **doi.org/10.3389/frobt.2021.781985**

Fink, W., Fuhrman, C., Zuniga, A. N., Tarbell, M., 2023, 'A Hansel & Gretel breadcrumb-style dynamically deployed communication network paradigm using mesh topology for planetary subsurface exploration', *Advances in Space Research*, **72**, #2, 518–528.

Flinders, L., Grainger, J., Clark, H., Rich, B. J., Braddock, M., 2022, 'Design architecture for 3D printing a Lunar habitat', *ROOM The Space Journal of Asgardia*, #1, 90–97.

Gordon, J-S., Pasvenskiene, A., 2021, 'Human rights for robots? A literature review', *AI and Ethics*, **1**, 579–591.

Hong, S., Bangunharcana, A., Park, J-M., Choi, M., Shin, H. S., 2021, 'Visual SLAM-Based Robotic Mapping Method for Planetary Construction', *Sensors*, 21, #22, **doi.org/10.3390/s21227715**

Jung, Y., Hahn, S., 2023, 'Social Robots As Companions for Lonely Hearts: The Role of Anthropomorphism and Robot Appearance', **arxiv.org/abs/2306.02694**

7. azquotes.com/quote/1567678

Mori, M., Translated by MacDorman K. F., Kageki Norri. 'The uncanny valley [From the Field]', *IEEE Robotics and Automation Magazine*, **19**, #2 (June 2012), 98–100.

Neta, S., Ariel, G., Yossi, Y., Amir, A., Ben, M. M., 2023, 'The Locust antenna as an odor discriminator', *Biosensors and Bioelectronics*, **221**, Article Number 114919.

Petráček, P., Krátký, V., Petrlík, M., Báča, T., Kratochvíl, R., Saska, M., 2021, 'Large-Scale Exploration of Cave Environments by Unmanned Aerial Vehicles', *IEEE Robotics and Automation Magazine*, **6**, #4, 7596–7603.

Qiu, W., Zhang, C., Chen, G., Zhu, H., Zhang, Q., Zhu, S., 2021, 'Colorimetric Ionic Organohydrogels Mimicking Human Skin for Mechanical Stimuli Sensing and Injury Visualization', *ACS Applied Materials & Interfaces*, **13**, #22, 26490–26497, **pubs.acs.org/doi/10.1021/acsami.1c04911**

Rahman, M. A., Walia, S., Naznee, S., Taha, M., Nırantar, S., Rahman, F., Bhaskaran, M., Sriram, S., 2020, 'Artificial Somatosensors: Feedback Receptors for Electronic Skins', *Advanced Intelligent Systems*, **2**, #11, **doi.org/10.1002/aisy.202000094**

Ribeiro, W. F. R., Uno, K., Imai, M., Murase, K., Yalçın, B. C., Hariry, M. E., Olivares-Mendez, M. A., Yoshida, K., 2023, 'Mobility Strategy of Multi-Limbed Climbing Robots for Asteroid Exploration', **arxiv.org/abs/2306.07688**

Sayed, M. E., Roberts, J. O., Donaldson, K., Mahon, S. T., Iqbal, F., Li B, Aixela, S. F., Mastorakis, G., Jonasson, E. T., Nemitz, M. P., Bernardini, S., Stokes, A. A., 2022, 'Modular Robots for Enabling Operations in Unstructured Extreme Environments', *Advanced Intelligent Systems*, **4**, #5, **doi.org/10.1002/aisy.202000227**

Shumeiko, V., Paltiel, Y., Bisker, G., Hayouka, Z., Shoseyov, O., 'A nanoscale paper-based near-infrared optical nose (NIRON)', *Biosensors and Bioelectronics*, **172**, (January 2021), **doi.org/10.1016/j.bios.2020.112763**

Takahashi, C., Giuliani, M., Lennox, B., Hamel, W. R., Stolkin, R., Semini, C., eds., 'Robotics in Extreme Environments', *Frontiers in Robotics and AI*, **8**, (2021), **doi.org/10.3389/978-2-88971-425-4**

Tootell, R. B. H., Zapetis, S. L., Babadi, B., Nasiriavanaki, Z., Hughes, D. E., Mueser, K., Otto, M., Pace-Schott, E., Holt, D. J., 2021, 'Psychological and physiological evidence for an initial 'rough sketch' calculation of personal space', *Scientific Reports*, **11**, Article Number 20960, **nature.com/articles/s41598-021-99578-1**

Eta Carinae
A Chance Encounter and Journey of Discovery

Peter Rea

Introduction

In 2004 my wife, daughter and myself were having a holiday in New Zealand. One cloudless evening we drove out into the countryside in search of dark skies. We soon found a suitable site and proceeded to set up my very basic equipment. I only had a Pentax film camera (I used Kodak Ektachrome 400 film), tripod and cable release. I could see the Small and Large Magellanic Clouds quite easily and the southern Milky Way stood out well. I pointed the camera by sight, rather than through the eye piece, toward a bright area of the Milky Way. Not having a driven mount I was advised to set exposures of no longer than 30 seconds. I took a series

Author's own image of the southern Milky Way. The Eta Carinae Nebula is the pink patch slightly below and to the right of centre. The Southern Cross (Crux) can be seen toward the upper left. (Peter Rea)

of frames along the Milky Way and both Magellanic Clouds. One of these frames revealed a pink patch, which my good friend and well known amateur astronomer Paul Money informed me was the Eta Carinae Nebula. I did not know what that was, so set about finding out. Some of what I discovered is included here.

Location, Size and Distance

A good starting point to track down the Eta Carinae nebula is to use the constellation of Orion, which straddles the celestial equator and can be observed from both the northern and southern hemispheres. Extending a line from Alnitak (the easternmost star in Orion's Belt) roughly southwards through Saiph (the star which marks the south-eastern corner of Orion) will eventually bring you to the brilliant Canopus. This is the brightest star in Carina, the constellation in which we find the Eta Carinae Nebula. The nebula itself is 500 light years away and 460 light years from one edge to the other, though sources vary slightly. At that size and distance the nebula appears about 120 arc minutes in the sky. If we could magically move the nebula to where the Orion Nebula is, then it would appear about twice the size of the Orion Nebula. It would be spectacular.

Lying some 7,500 light years away is the Eta Carinae Nebula hosting many O-type and Wolf-Rayet stars, numbered amongst which is the double star Eta Carinae. (Wikimedia Commons/ Klausdonath)

Constellation

In 1763, the French astronomer Nicolas-Louis de LaCaille published his *Coelum Australe Stelliferum* or "Southern Starry Sky". This work listed many of the southern constellations, including Argo Navis which was one of the 48 constellations known to Greek astronomers and catalogued by Ptolemy in his famous book now known by its Arabic name *Almagest*. LaCaille noted that Argo Navis – originally named after the ship Argo which was sailed by Jason and the Argonauts during his search for the Golden Fleece – was a large and sprawling constellation. He proceeded to divide it into three separate constellations, these being

The constellation Argo Navis drawn by Johannes Hevelius and appearing in the catalogue *Uranographia*. This sprawling constellation was later split into three separate constellations: Vela, Puppis and Carina. (Wikimedia Commons / Johannes Hevelius)

Vela (the Sails), Puppis (the Poop Deck or Stern) and Carina (the Keel). Of the 48 constellations listed by Ptolemy, Argo Navis is the only one no longer recognised. As a result of the dismantling of Argo Navis, the star previously known as Eta (η) Argûs became Eta Carinae. The nebulosity surrounding this double star is referred to as the Eta Carinae Nebula, or NGC 3372.

Great Eruption

The star Eta Carinae was barely visible to the naked eye during the eighteenth and early-nineteenth centuries, although by the 1830s it had brightened to fourth magnitude. In 1837 it suddenly brightened to magnitude −1 and equalled the star Canopus (Alpha Carinae). This brightening is known as "The Great Eruption." By 1856 it had once again fallen to below naked eye visibility.

Sir John Herschel Observations

In November 1833 Sir John Herschel, only son of Sir William Herschel and Mary (nee Pitt), set off for the Cape of Good Hope aboard the sailing ship *Mount Stewart Elphinstone* accompanied by his wife Margaret and three of their children. Amongst the items on board was his father's 20-foot telescope, originally sighted at Datchet and Slough. On arrival in South Africa he moved his family to the Grove Estate, located in the Claremont district of Cape Town and which he bought and renamed

Engraving depicting the site of the Twenty Feet Reflector at Feldhausen in September 1834. Table Mountain is seen in the background, and Cape Town would be on the other side of the mountain at the two o'clock position. The smaller structure to the right of the image houses a 5-inch refracting telescope. (Wikimedia Commons/ John F. W. Herschel/George Henry Ford/ Smith, Elder & Co)

Feldhausen. His intention was to carry out a major survey of the southern sky, in the same way that his father had mapped the northern sky, and in an effort to "... finish my father's work." Among many observations recorded later in his catalogue *Results of Astronomical Observations Made During The Years 1834, 1835, 1836, 1837, 1838, At The Cape of Good Hope* were observations of the Great Eruption and drawings of the Eta Carinae Nebula, all achieved with the 20-foot telescope. One of many drawings of this region shows a feature known as The Keyhole Nebula. Sir John was an excellent artist, as was his wife Margaret whose skills are revealed by her many drawings of South African flora.

Homunculus Nebula

Today we can still see evidence of that Great Eruption. Astronomers who turn their telescopes toward Eta Carinae will observe an expanding cloud of material known as the Homunculus Nebula. In the nearly-two-centuries since the brightening started, the two expanding lobes have expanded to a total of around one light year across, images taken just a few years apart revealing this expansion. Buried within

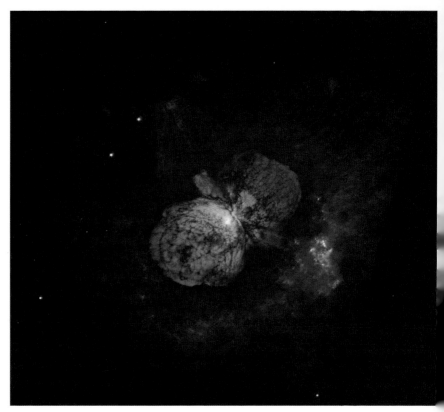

Image of the Homunculus Nebula, captured by the Hubble Space Telescope. The nebula was caused in 1837 by the Great Eruption of the double star Eta Carinae, and is still expanding (Nathan Smith (University of California, Berkeley), and NASA)

this expanding cloud is the star Eta Carinae, which we now know to be a double Both stars are very large – and very hot – O-type stars, orbiting each other over a period of 5.5 years. At periastron (the point in their orbits at which they are closest to each other) the two stars approach to within a distance of 224 million kilometres roughly equal to the distance from our Sun to the planet Mars. The larger of the two stars is around 100 solar masses, the smaller star being around 30 solar masses although sources vary and these are only approximations. What everyone appears to be in agreement with is that, in the near future – astronomically speaking – both stars will collide in a massive explosion. The aftermath of this event would be a great sight to southern hemisphere observers, those dwelling in the northern hemisphere having to settle for pictures!

Saturn at its Equinox
A History of Ring-Plane Crossings from 1612 to 2025

David Harper

The rings of Saturn have been a source of wonder and delight to astronomers for more than four centuries. Their constantly-changing aspect makes them a fascinating sight. But twice in every Saturnian year (29.4 Earth years), they vanish completely as the Sun and Earth pass through the plane of the rings. In this article, we explore the history of ring-plane crossings from the first observed event in 1612 to this year's disappearance.

Vanishing once, twice, thrice

The Sun crosses the ring-plane twice in each Saturnian year. These are Saturn's equinoxes, when the Sun stands directly over the planet's equator. Around the time of each Saturnian equinox, the Earth may pass through the ring-plane just once, or it may cross the rings three times. A single crossing occurs when Saturn is close to conjunction with the Sun. This is what happens in 2025: the Earth passes through the ring-plane on 23 March, when Saturn is 11 days past conjunction and only 10° from the Sun. In 2038/39, by contrast, the Earth will pass through the ring-plane three times, and the second crossing, on 1 April 2039, is just two week after opposition. In 1908, the Oxford astronomer Herbert Turner showed that triple crossings are slightly more likely (53% of equinoxes) than single crossings (47% of equinoxes).

Seventeenth and eighteenth century ring-plane crossings

The first ring-plane crossing to be observed was that of 1612/13, just three years after Thomas Harriot and Galileo Galilei ushered in the era of telescopic astronomy. Galileo, observing Saturn in 1610, mistook the rings for large moons because his telescope did not have the resolving power to reveal their true nature. He was astonished when they vanished two years later. By 1616, they had opened up sufficiently for Galileo to sketch them as two half-ellipses on either side of the planet.

Saturn's rings are composed predominantly of water ice, reflecting as much as 60% of the incident sunlight. They contribute significantly to Saturn's overall brightness, making it a full magnitude brighter at opposition when the rings are

On 26 July 2009, NASA's Cassini spacecraft captured this image of the outer edge of Saturn's B ring. The Sun casts long shadows of vertical structures which tower 2.5 kilometres above the plane of the rings. (NASA/JPL/Space Science Institute)

fully open than when they are edge-on. From the seventeenth century to the era of the Hubble Space Telescope, ring-plane crossings have proven to be a very fruitful time to look for new moons, as the overwhelming brightness of the rings is temporarily abated.

At the triple-crossing of 1655/56, the Dutch astronomer Christiaan Huygens discovered Titan, the largest and brightest moon of Saturn, and hypothesised that the planet was surrounded by a thin, flat ring. Like most of his contemporaries, he assumed that the ring must be solid.

The next ring-plane crossing, in 1671/72, was also a triple event for the Earth. The Italian astronomer Jean-Dominique Cassini, working at the royal observatory in Paris, discovered first Iapetus and then Rhea in the late autumn and early winter of 1671. Bad weather and Saturn's proximity to the Sun prevented him from continuing his observations until the autumn of 1672, when the tilt of the rings towards the Earth was barely 3°. Using a newly-acquired and more powerful telescope, Cassini recovered Iapetus and correctly deduced its orbital period to be around 80 days.

William Herschel's 48-inch reflector, build during the late 1780s, was the largest telescope in the world for half a century. Herschel used this telescope to study Saturn during the 1789 ring-plane crossing. (Leisure Hour (2 Nov 1867)/Wikimedia Commons)

The ring-plane crossing of August 1685 occurred close to conjunction, but the tilt of the rings towards Earth dropped below 1° during December 1684, enabling Cassini to discover two new moons, Tethys and Dione.

More than a century would pass before Saturn's family of moons was expanded again. William Herschel took advantage of the triple-crossing of 1789/90 to discover Mimas and to confirm his earlier discovery (in August 1787) of Enceladus. Herschel observed several of the moons passing behind the rings on nights close to the Earth's crossing of the ring-plane on 27 August 1789, and noted that he could see each moon bisected by the almost edge-on ring. However, he hesitated to infer from this that the rings must be extremely thin.

Observations of the "dark" ring in the nineteenth and early-twentieth centuries

During the nineteenth century, interest in Saturn's rings turned to their structure and composition. In 1857, the Scottish mathematician James Clerk Maxwell submitted an essay for Cambridge University's prestigious Adams Prize. His subject was the nature of the rings of Saturn, and he proved that the rings could neither be solid nor fluid and remain stable. He concluded that the rings must be composed of innumerable small particles, each orbiting Saturn independently.

A number of now-famous astronomers made observations of Saturn's edge-on rings in the nineteenth century, including William Bond and his son George at the Harvard College Observatory in 1848 (when they also discovered Hyperion), James Carpenter (at Greenwich) and Otto Struve (at Pulkova in Russia) in 1861. Both Carpenter and Struve, using 12- and 15-inch refractors respectively, were able to see the rings when they were "unlit", with the Earth and Sun on opposite sides of the ring plane.

The 1878 and 1891 ring-plane crossings both occurred close to conjunction, and the next opportunity to observe the "dark" rings came in 1907/08, another triple-crossing. The American astronomer Edward Emerson Barnard used the 40-inch refractor at the Yerkes Observatory to observe the rings in July 1907 at the Sun's crossing of the ring plane, at the Earth's second crossing in October 1907, and for the remainder of 1907, when the Earth and Sun were on opposite sides of the ring.

Barnard's 1908 report of his observations includes many fascinating details, together with a sketch which he made of the back-lit rings on 12 December 1907 which shows two distinct bright zones on either side of the planet. He identified these "condensations" with the C (or "crêpe") ring and the outer edge of the B ring adjacent to the Cassini division. He then discussed the meaning of these observations. He noted that in November 1889, the satellite Iapetus had been

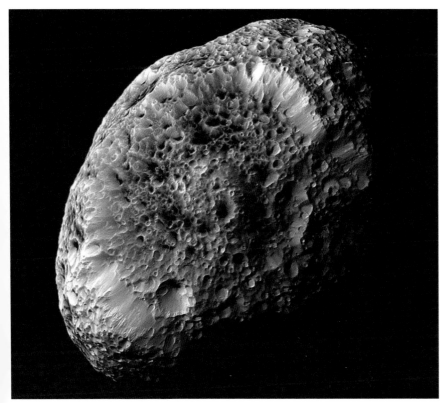

Saturn's moon Hyperion was discovered during the 1848 ring-plane crossing by William and George Bond at Harvard and, independently, by William Lassell in Liverpool. This Cassini image reveals its very peculiar appearance. (NASA/JPL/Space Science Institute)

eclipsed by the shadow of the ring system, disappearing completely when in the shadow of the A and B rings but remaining visible (albeit dimmed by about a magnitude) whilst in the shadow of the C ring. Thus, he concluded, the C ring must be partly transparent. The bright "condensation" which he observed in 1907, he suggested, was due to sunlight reflected at an oblique angle by the myriad of small particles that make up the ring.

Co-orbiting Moons – the ring-plane crossing of 1966
In December 1966, as the Earth passed through the ring-plane for the third time in the triple-crossing of that year, the French astronomer Audouin Dollfus identified a new, faint (magnitude 14) moon in photographic plates obtained on

In this mosaic of images obtained by NASA's Cassini spacecraft on 12 August 2009, just 1.25 days after Saturn's equinox, the rings are illuminated by sunlight reflected from Saturn itself. The co-orbital moons Janus and Epimethus can be seen just outside the rings, at the 9 o'clock and 6 o'clock positions. (NASA/JPL/Space Science Institute)

three successive nights. This moon orbited just beyond the outer edge of the ring system. American astronomer Richard Walker, working at the Flagstaff station of the U.S. Naval Observatory, also captured an image of a faint satellite orbiting at the same distance, and assumed it was the same object. Twelve years later, John Fountain and Stephen Larson of the Lunar and Planetary Laboratory, University of Arizona, re-analysed the 1966 observations and concluded that they could not be explained by a single satellite. There must be two moons in almost identical orbits. Dollfus is now credited with the discovery of Janus, and Walker with the discovery of Epimetheus.

The existence of two moons was confirmed by *Voyager 1* when it visited Saturn in November 1980 and obtained images of both objects. They are irregular, with mean diameters of 175 kilometres and 105 kilometres, yet their orbits are only separated by 50 kilometres in distance from Saturn. This suggests that they must inevitably collide with one another within just a few years, but there was no evidence of past collisions. The solution to this puzzle was provided by Stanley Dermott and Carl Murray, working at the University of Florida in 1981. They demonstrated mathematically that the two moons would approach one another no closer than about 10,000 kilometres and effectively exchange orbits, with the inner moon moving to the outer orbit and vice versa. After this exchange, the moons would begin to move apart, meeting again every four years to repeat the process.

The Hubble Space Telescope and the ring-plane crossing of 1995

The 1995 ring-plane crossing was another triple event for the Earth, whose 22 May and 10 August crossings were well-placed for observers, as was the Sun's crossing on 19 November. Astronomical technology had progressed significantly since 1966. Ground-based observers could use CCD cameras and adaptive optics, whilst the Hubble Space Telescope (HST) offered views of unrivalled clarity from above the Earth's atmosphere.

A major observing campaign began in May 1995, as the rings became edge-on to the Earth. A team led by Amanda Bosh at the Lowell Observatory in Arizona used the HST to perform photometry of the rings in the hours either side of the predicted moment when the rings should vanish. They found that the Earth crossed the ring plane somewhat later than expected, and from this, they deduced that the rate of precession of Saturn's pole must be slower than had been thought, or that the assumed orientation of the pole was not correct. They also determined the thickness of the rings: a mere 1.4 kilometres. This figure was subsequently confirmed by another team, led by Phil Nicholson of Cornell University, who used the HST to observe Saturn at the 10 August Earth crossing and the period from 17 to 21 November when the Sun crossed the ring plane.

The Cassini Spacecraft and its Equinox Mission

The 2009 ring-plane crossing was very unfavourable for Earth-based observers, occurring in August–September of that year, when Saturn was very close to conjunction.

However, NASA's *Cassini* spacecraft was then in orbit around Saturn, providing astronomers with the first view of a ring-plane crossing from the planet itself. The spacecraft had arrived at Saturn in July 2004 at the end of a seven-year journey. The nominal length of the orbital tour was four years, but in early 2008, with the spacecraft still operating almost flawlessly, NASA approved a 27-month extension. This was named the *Cassini Equinox Mission*, as it would include the 2009 ring-plane crossing.

On 26 July 2009, with the Sun less than a quarter of a degree below the ring-plane, Cassini's narrow-angle camera captured a remarkable image of the outer edge of the B ring. Vertical structures tower as much as 2.5 kilometres above the plane of the rings. Shining brilliant white, they cast long, jagged shadows across the rings.

Orbiting the planet every 80 days, Cassini crossed the ring-plane many times during its 13-year orbital tour, and it obtained numerous images with the rings edge-on. This enabled astronomers to determine that the main A, B and C rings are no more than about 10 metres thick.

NASA's Cassini spacecraft crossed the ring-plane many times during its 13-year orbital tour. In this image, taken on 17 August 2017, the extreme thinness of the rings is evident. The moon Dione (diameter 1,123 kilometres) adds a sense of perspective. (NASA/JPL/Space Science Institute)

The 2025 ring-plane crossing

The Earth crosses the ring-plane from north to south on 23 March 2025, when Saturn is only 10° west of the Sun, so this event is invisible from the Earth. Saturn's equinox follows on 6 May, with Saturn 48° from the Sun in the morning sky. Please refer to the article *Saturn at its Equinox: The Ring Plane Crossing of 2025* following the May sky notes for a detailed description of the observational prospects at this ring-plane crossing.

Looking forward to 2038/39

The next ring-plane crossing, in 2038/39, is a much more favourable event. The Earth crosses the ring-plane three times. On the first occasion, Saturn is too close to the Sun to be seen, but the interval between the Earth's second crossing on 1 April 2039 (16 days after opposition) and its third crossing on 9 July 2039 (Saturn 67° East of the Sun) provides an extended opportunity to observe Saturn with its rings unlit, tilted by almost a full degree during most of May 2039.

Selected References

Barnard, E.E. (1908) "Observations of Saturn's rings at their disappearances in 1907 with a suggested explanation of the phenomena presented". *Astrophysical Journal*, **27**, 35–44. **articles.adsabs.harvard.edu/pdf/1908ApJ....27...35B**

Cassini, J.-D. (1672) "A Discovery of Two New Planets about Saturn, Made in the Royal Parisian Observatory by Signor Cassini, Fellow of Both the Royal Societys, of England and France; English't Out of French". *Philosophical Transactions of the Royal Society of London*, **8**, 5178–5185. **jstor.org/stable/101334**

Cassini, J.-D. (1673) "Découverte de deux nouvelles planètes autour de Saturne". Bibliothèque numérique – Observatoire de Paris. **bibnum.obspm.fr/ark:/11287/2z73h**

Herschel, W. (1790) "Account of the Discovery of a Sixth and Seventh Satellite of the Planet Saturn; With Remarks on the Construction of Its Ring, Its Atmosphere, Its Rotation on an Axis, and Its Spheroidal Figure". *Philosophical Transactions of the Royal Society of London*, **80**, 1–20. **jstor.org/stable/106823**

Herschel, W. (1790) "On the Satellites of the Planet Saturn, and the Rotation of Its Ring on an Axis". *Philosophical Transactions of the Royal Society of London*, **80**, 427–495. **jstor.org/stable/106848**

Turner, H.H. (1908) "Saturn, note on the condition for the passage of the Earth through the plane of the ring". *Monthly Notices of the Royal Astronomical Society*, **68**, 460–464. **articles.adsabs.harvard.edu/pdf/1908MNRAS..68..460T**

A History of Observatory Designs
Before the Telescope

Katrin Raynor

Prior to the invention of the telescope in (what is believed to be) 1608 by Dutch lens maker Hans Lippershey, our ancestors looked to the night sky without the use of optical aids. The naked eye was their only means of observing the stars and planets. Imagine how incredible this would have been, with the sky as black as ink and the heavens above thick with stars with no interference from light pollution. The view we had then is very different to what we have now.

There is global evidence that early humans studied the skies to observe lunar and solar events such as eclipses and the solstices. The observations made were used for farming, to regulate calendars and for ritualistic purposes. During prehistoric times, observatories generally took the form of stones arranged in alignment to the rising and setting of the Sun and Moon and for tracking the seasonal constellations. Often referred to as observing posts, for the purpose of this article they will be specified as observatories. It was to be thousands of years before professional observatories were constructed to make detailed calculations of the heavens.

In this and forthcoming editions of the *Yearbook of Astronomy*, we will be exploring observatory design from prehistory to the modern period. The following article features just a handful of examples, with brief descriptions, of observatories dating from prehistory through to the period immediately preceding the invention of the telescope.

Stonehenge

The United Kingdom is rich in prehistoric archaeology, including artefacts, hill forts, and burial sites, providing an insight to how early humans lived and interacted with each other and the surrounding land. Perhaps the most famous of all prehistoric archaeological sites in Europe is Stonehenge in Wiltshire, a UNESCO World Heritage site. Built over 5,000 years ago during the Neolithic period and early Bronze Age, Stonehenge is thought to have been completed in three phases over a 1,500-year time frame. Its circular structure is unmistakable due to the construction of the outer vertical sarsen standing stones connected by horizontal lintel stones placed on top. Outside of the circle is the Heel Stone. Sometimes referred to as

One theory surrounding the Station Stones at Stonehenge is that there were once four of them forming a rectangle shape with the centre of this rectangle being at the midpoint of the stone monument. These Station Stones may have been used to measure variations in the orbital plane of the 18.6-year lunar nodal cycle. On 5 February 2023, Nick Bull captured this beautiful image of a rising Snow Moon behind Stonehenge, igniting our imaginations of what lunar events like this must have looked like thousands of years ago. (Nick Bull/ Stonehenge Dronescapes Photography/**facebook.com/stonehengedronescapes**)

'The Friar's Heel', this huge and isolated sarsen boulder[1] is surrounded by a ditch, and a hole located next to it which may have been the site of a second Heel Stone.

For hundreds of years antiquarians and archaeologists helped shape our understanding of the site and its use. John Aubrey (1626–1697) and William Stukeley (1687–1765) performed the first surveys and excavations during the decades following 1670. However, it was not until the late-eighteenth century that the first hypothesis of Stonehenge as a solar observatory was proposed by James Douglas (1753–1819). A century later scientist, astronomer and pioneer in archaeoastronomy Sir Norman Lockyer (1836–1920) attempted to date Stonehenge using the Heel Stone. Assuming its orientation to sunrise at mid-summer, he concluded that the building of the site dated back to 1680 BCE. In 1952, radiocarbon dating of the stones calculated a date of 1800 BCE.

Research is detailed and complex with plenty of reading material available around the theories of Stonehenge. One (out of many) of the most accepted theories is its

---

1. Information about the silicified sandstone blocks known as sarsen stones can be found at **wikipedia.org/wiki/Sarsen**

C/2020 F3 NEOWISE or Comet NEOWISE, delighted astronomers in 2020 when it became bright enough to be seen with the naked eye. This stunning image of NEOWISE gliding above Stonehenge was taken by Nick Bull during the early morning of 17 July 2020. Many theories around the design of Stonehenge focus on its use as an observatory for observations of eclipses, solstices, and cycles of the Moon. Imagine how our ancient ancestors would have felt, seeing a comet in the skies above the stones with no understanding whatsoever of what this silent heavenly visitor was. (Nick Bull/ Stonehenge Dronescapes Photography/**facebook.com/ stonehengedronescapes**)

use as an astronomical observatory and that its axial alignment is to the summer and winter solstices. When standing in the centre of the stone circle and looking towards the north east at the rising Sun on the summer solstice in June, the Sun can be seen rising over the solitary Heel Stone. It is thought that there were in fact two Heel Stones, and that the rising Sun was framed by the pair. Similarly, on the afternoon of the winter solstice, if an observer were to stand at the Heel Stone and look towards the vertical lintels, the Sun can be seen setting between the stones.

Stonehenge may not have been just a solar observatory. The presence of Station Stones at the monument is thought to have been used to record lunar alignment. Although there are only two Station Stones remaining, there is evidence to suggest that there were once four which together marked the corners of a rectangle, the central point of which was located at the exact midpoint of the monument. The carefully placed stones forming the rectangle may have been used to measure the 18.6-year lunar nodal cycle of variations in the Moon's orbital plane.

Bryn Celli Ddu

Bryn Celli Ddu (the 'Mound in the Dark Grove') on the island of Anglesey in north Wales is one of Anglesey's most famous prehistoric landmarks and a beautiful example of an early Neolithic henge (bank and ditch) and burial chamber. Constructed around 5,000 years ago, it is perfectly aligned to the dawn of the summer solstice. The henge consisted of a bank sitting within an inner ditch, a

Located on the Island of Anglesey, Bryn Celli Ddu, or the 'Mound in the Dark Grove', is one of Wales's best examples of a Neolithic burial chamber. Excavated over a year between 1928 and 1929, studies of the site showed that it was originally a henge, but around a thousand years after its original construction it had been transformed into a burial chamber. Sunlight penetrates the passageway on the summer solstice, illuminating the burial chamber inside. (Wikimedia Commons/Jeff Buck/Attribution-ShareAlike 2.0 License)

monument of seventeen stones were constructed and enclosed within. Around a thousand years following its original construction, the site was altered into a passage chamber, a form of burial monument with a long passageway leading to a polygonal shaped chamber. Remarkably, the positioning of the burial chamber and passageway is so accurate, that on the dawn of the summer solstice, sunlight penetrates the passageway to illuminate the burial chamber inside.

Ulugh Beg Observatory

Constructed in the 1420s, the Ulugh Beg Observatory in Samarkand, Uzbekistan was built by the Timurid sultan, astronomer and mathematician Mīrzā Muhammad Tāraghay bin Shāhrukh, better known as Ulugh Beg (1394–1449). Beg's interest in astronomy was aroused after visiting the remains of the Maragheh observatory, established during the mid-thirteenth century in Persia (now Iran). Although Maragheh observatory had become inactive by the early-fourteenth century, its design influenced a number of other observatories, including the one constructed by Ulugh Beg.

In an effort to make Samarkand a centre for intellectuals, Ulugh Beg constructed a university for mathematicians and astronomers there, as well as the biggest observatory of its time in Central Asia. Notable astronomers who visited Beg and studied at the observatory include Jamshid al-Kashi and Qadi Zada. (Wikimedia Commons/Ulugh Beg Observatory/BerkBerk68)

Round, with a diameter of 46 metres and split over three floors, the observatory housed the main instrument, a forty-foot radial sextant. To aid stability, part of this enormous astronomical tool was embedded in a trench dug into a hill. According to the Portal to the Heritage of Astronomy,[2] the size of the sextant allowed for very accurate graduation. One degree on the arc of the sextant was represented by a division of 70.2 centimetres with marks 11.7 millimetres and one millimetre apart representing one minute and five seconds respectively.

The sextant proved to be very successful, permitting accurate measuring of the Sun on the horizon and the altitudes of planets and stars. It was also used

2. Portal to the Heritage of Astronomy: **astronomicalheritage.net**

Taken in 2001, this image of the Ulugh Beg Observatory shows the main entrance, which has undergone numerous aesthetic changes over the years. The observatory was active for around 30 years before it was destroyed in 1449, but in 1908 the Russian archaeologist Vassily Vyatkin rediscovered the site and unearthed its remains. (Wikimedia Commons/Michel Benoist/GNU Free Documentation License)

to measure eclipses, planetary motion, and the length of a year, with one of its most notable successes – achieved with the aid of other astronomers – was the calculation of the obliquity of the ecliptic.

Beg's goal was to write and publish a *Zij*, an Islamic astronomical book containing tables of parameters to calculate positions of the Moon, Sun, planets and stars. Taking twenty years to write, Beg successfully published the *Zīj-i Sulṭānī* in 1438–1439. It contained a list of the positions of just over a thousand stars – including 27 from the Persian astronomer Abd al-Rahman al-Sufi's *Book of Fixed Stars*, published in or around 964 CE – as well as accurate calculations of the length of one solar year and the axial tilt of the Earth.

As was the case with many ancient observatories, Beg's observatory was destroyed (in 1449), with only eleven metres of the sextant now remaining. A museum now stands at the site and includes a large statue commemorating Beg.

Following the archaeological dig by Vassily Vyatkin, all that remains are foundations of the observatory and part of the huge sextant constructed by Beg. Due to its enormous size, Beg built half of the sextant underground, to prevent having to construct a very tall and potentially unstable observatory to accommodate the massive instrument. The sextant was oriented in the meridian plane and used to measure, for example, the angle of elevation of the planets. The walls seen in the photograph would originally have been lined with marble tiles. (Wikimedia Commons/Igor Pinigin/GNU Free Documentation License)

Uraniborg

Tycho Brahe (1546–1601) is one of the best known and celebrated astronomers in history. Born in Demark, Brahe was not only an astronomer, but an alchemist and astrologer famed for his accurate astronomical observations. He was also the last eminent astronomer to make observations and calculations before the development of the telescope.

Brahe rose to fame in early-November 1572 when he observed a 'new star' in the constellation of Cassiopeia. Now known as Tycho's Supernova, the object remained at naked-eye visibility until early-1574, and even became visible in daylight for a short time. Brahe's accurate calculations proved that this 'star' contradicted the

This hand-coloured copper-plate engraving of Tycho Brahe's Uraniborg observatory appeared in Volume 1 of the Dutch cartographer and publisher Joan Blaeu's *Atlas Maior*, published in Amsterdam between 1662 and 1665. The illustration is based on a woodcut which appeared in Brahe's own 1598 book *Astronomiæ instauratæ mechanica*. Tycho Brahe's Uraniborg observatory was the first to be custom built on mainland Europe, and the last to be built without the presence of a conventional telescope. Uraniborg observatory was built in the Flemish Renaissance style and named after *Urania*, the Muse of Astronomy in Greek mythology. Brahe's observatory was destroyed in 1601, shortly after his death. (Wikimedia Commons/Tycho Brahe/Willem Blaeu/Joan Blaeu/National Library of Scotland)

Another hand-coloured copperplate engraving taken from Joan Blaeu's *Atlas Maior*, this one depicting a front view of the square, three storey main building, along with its towers and balconies, of Tycho Brahe's Uraniborg observatory. It was from here that Brahe and his assistants carried out most of their astronomical observations – including studies of the Great Comet of 1577 – from the roof, while the cellar served as a laboratory for alchemy. (Wikimedia Commons / Tycho Brahe / Willem Blaeu / Joan Blaeu / National Library of Scotland)

Aristotelian doctrine that the cosmic realm beyond the Moon was unchanging He concluded that the object was located far beyond the Moon, leading him to the conclusion that it was possible for the universe to undergo changes.

In 1576, King Frederick II of Denmark granted Brahe an estate on the island of Hven – situated in the Øresund (the 'Sound') between Zealand and Scania, Sweden and which was part of Denmark at the time – to build an observatory. This was the first observatory to be custom built on mainland Europe and the last without the presence of a telescope. It was named Uraniborg after *Urania*, the muse of astronomy, and was a place where Brahe felt very much at home and where he and his assistants spent around twenty years building instruments and undertaking research on astronomy, astrology, alchemy, and meteorology.

The impressive Uraniborg observatory took approximately four years to build, the construction being overseen by the Flemish-Danish architect Hans van

Steenwinckel der Ältere and the Northern Renaissance sculptor Johan Gregor van der Schardt, in partnership with Tycho. The main building was located inside a beautiful Renaissance garden of symmetrically patterned plants and hedges, gravel and paths. Uraniborg was square in shape, measuring fifteen metres on each side and erected using a frame of sandstone and limestone.

The observatory building had one large tower standing at 19 metres tall, together with two smaller towers located at the north and south sides. Brahe's instruments included quadrants, sextants and armillary spheres, all of which were housed in the towers and on the balconies from where observations of the night sky were carried out. Within the observatory, and mounted on a wall that was oriented precisely north-south, was a large mural quadrant which was used to measure the altitude of stars as they passed the meridian. The cellar served as an alchemy laboratory for Brahe.

During his time at Uraniborg, Brahe's achievements were unprecedented. He observed the comet of 1577 which led him to develop the Tychonic System, a geoheliocentric model and made accurate charts of the stars and planets. Tycho also developed new methods for observing, moving away from the traditional construction of instruments using wood and turning his hand to masonry and metal work to improve instrument durability. Instruments and equipment were affixed to the observatory walls in order to increase their stability – and therefore improve their accuracy – whilst at the same time sheltering them from the elements.

Shortly after Brahe's death in 1601 the observatory at Uraniborg was destroyed, along with many of Tycho's instruments, although his meticulously compiled written records provide us with a fantastic and revealing insight into his research and the observatory and instruments he created. Excavation and restoration work of Uraniborg commenced during the 1950s and this now forms part of the Tycho Brahe Museum.

Further Information

Information relating to Stonehenge and other ancient henge sites and stone circles located within the UK can be found at **stonehenge.co.uk**

Bryn Celli Ddu: **cadw.gov.wales/visit/places-to-visit/bryn-celli-ddu-burial-chamber**

Uraniborg Museum: **visithven.dk/en/tycho-brahe-museum**

Signals from the Magnetosphere

John Vetterlein

The Merry Dancers

Most Orcadians[1] are familiar with the appearance in the night sky from time to time of what we call locally the *Merry Dancers*. The description is apt in many instances where auroral elements are seen to literally dance across the sky. However, there are other auroral features that do not fall into this category, including the often all-pervading north *auroral glow* and *auroral arcs*. The area of sky most affected is generally north of the zenith and, in the case of glows and arcs, fairly close to the northern horizon itself.

What Are the Aurora and Where Do They Occur?

It will be seen from the diagram that aurora may occur at heights above the Earth's surface of between 80 kilometres and 250 kilometres. At these altitudes, the density of the atmosphere – and comprised mostly of molecules and atoms of nitrogen and oxygen – is very low. We now recognize that particles from within the solar wind (atomic sub-particles – protons and electrons) and originating from the Sun itself are attracted by the Earth's magnetic field. This leads to them becoming concentrated above the region of the Earth's magnetic poles. Here they collide with the atoms and molecules of nitrogen and oxygen, emitting light in various forms that we witness as the aurora. The Earth is being perpetually bombarded by the solar wind, the extent of the bombardment varying considerably over time and being greatest at times of high solar activity.

The Sun is an ordinary star, one of millions in our own Milky Way Galaxy. But for us here on Earth the Sun is immense importance. In general we take it for granted; the Sun rises and sets day after day as it has done for millions of years. Looking at it from this perspective, our detailed knowledge of the Sun is of very recent origin.

Coincidentally the correlation between solar activity and the aurora was only beginning to be understood around 160 years ago as a result of the violent magnetic

---

1. The ethnic group native to the Orkney Islands, also known as Orkneymen. For more information visit **wikipedia.org/wiki/Orcadians**

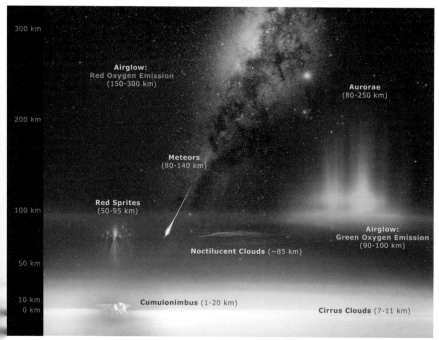

300 km

Airglow:
Red Oxygen Emission
(150-300 km)

Aurorae
(80-250 km)

200 km

Meteors
(80-140 km)

Red Sprites
(50-95 km)

100 km

Airglow:
Green Oxygen Emission
(90-100 km)

Noctilucent Clouds (~85 km)

50 km

10 km

Cumulonimbus (1-20 km)

0 km

Cirrus Clouds (7-11 km)

A cross-sectional representation of the Earth's atmosphere showing various phenomena that occur within it including the aurora, which appear at heights of between 80 kilometres and 250 kilometres above the Earth's surface. (P. Horálek /NOIRLab /NSF /AURA)

storm of 1/2 September 1859. On an almost worldwide scale, aurora appeared in great abundance and intensity; not only that but other phenomenon forced their attention upon us including erratic behaviour in compasses and magnetometers, discharges from electric cables and so on.

On 1 September 1859, the English amateur astronomer Richard Christopher Carrington (1826–1875) based at Redhill, Surrey, was drawing a large sunspot complex copied from the projected image formed on a screen by his 115mm refracting telescope, when he noted a bright feature tangling with the spots. It soon became evident that this was in fact a solar feature in its own right. He noted the movement of the filaments and calculated that their tangential velocity must be in the region of a phenomenal 100 kilometres per second. On the evening and night of 2 September vivid aurora were widely reported, together with other freak events described above.

The periodicity of the solar cycle between repeated highs and lows varies somewhat, occupying an interval in the order of 11 years. During peaks of increased activity we observe more sunspots on the solar surface, together with more solar flares and an upsurge in emitted radiation. At the time this article was written (early-2023) the sun was undergoing a period of noticeably increased activity. Whether or not this produces storms comparable to those experienced in 1859 remains to be seen, although in our modern world, so dependent upon micro-electronics, the consequences could be devastating.

At times of maximum solar activity sunspots may appear in profusion and reach a scale resulting in them becoming visible to the unaided eye. Long before the development of optics, observers would have been able to see these spots when the Sun appeared in mist or thin cloud. It should be stressed of course that on no account should the Sun ever be viewed directly without taking the necessary precautions – see below.

Typical Auroral Features

In the region of high latitudes beneath the *auroral ovals* the night sky is seldom dark, even during mid-winter. An auroral glow may at times swamp the faint light of the Milky Way itself. The glow is generally whitish in hue – some have likened it to a night *haar*, or sea mist, high in the sky.

Spikes appear from an intense northern auroral glow in this image, taken from Rousay, Orkney on the evening of 9 March 2018. The display went on to yield short-lived rays, which appeared at intervals well into the night, the auroral glow itself persisting until a combination of dawn, moonrise and gathering cloud intervened. The distinctive constellation of Lyra, along with the bright star Vega, can be seen at lower right of the image. (John Vetterlein)

In this image, taken shortly after midnight on the night of 27/28 September 2016, we see a series of active rays which went on to culminate in a fine display. The distinctive outline of the 'Keystone' in the constellation Hercules, along with the showpiece globular cluster Messier 13, can be seen at far left centre of image. (John Vetterlein)

When observing from Orkney and similar latitudes, an auroral glow may appear on almost any clear night. Sometimes the glow will rise above the northern horizon to elevations of 30 degrees or more. (Bear in mind magnetic north is approximately five degrees west of true or geographical north.) The glow may show arc banding looking like a white rainbow stretching from the north western to north eastern horizon; on occasions there may be two arcs, one above the other.

During periods of solar activity we can expect to see other forms. These often develop out of the glow, appearing as *spikes* that may quickly develop into *rayed arcs*. These may further develop into *bands* that reach up into the sky, perhaps as far as the zenith (see below) and beyond. The form of the band or beam can change rapidly – within seconds – broadening at times and changing in colour and intensity.

In times of high solar activity bands can spread out to form homogenous features which may appear as greenish "curtains" that coil and ripple across the northern sky. Large fan-like features often appear and may be of an intense brick-red or deep blue in colour.

One of the most interesting configurations is the so-called *corona* – a result of perspective – that appears when bands, in particular, appear close to the zenith. All these forms the writer has witnessed personally from Rousay within the past twenty five years, and some of which are shown in the accompanying images.

In extreme conditions the glow may appear to break up and to transmit flashing lights across the sky from north to south; on other occasions light, cloud-like patches may similarly traverse the sky in rapid succession. From the latitude of Orkney it is not uncommon during these extensive displays to find the entire sky lit by aurora.

Is It Possible That Aurora Emit Audible Sound?

Some observers have reported sounds they associate with the more dramatic auroral displays. For sound waves to be transmitted over distance requires a substantial medium – air below a certain density cannot do this effectively and sound cannot travel through a vacuum. For most of the distance between the Earth

The gentle arc of the auroral glow is captured on this image, looking due north and taken on the evening of 31 December 2016. Lasting for around eight hours, this particular display was persistent and intense at times, spanning around 170° in azimuth and rising to 38° altitude. The conspicuous outline of the Plough holds centre stage. (John Vetterlein)

The early evening of 14 December 2015 gave a sky of scattered, dense cloud with a brilliant auroral glow showing in the gaps, bands and rays with patches of red and green appearing to the north east. The constellation of Gemini occupies much of the lower part of this image, its leading stars Castor and Pollux prominent to the left of centre. (John Vetterlein)

and the Moon, for example, we encounter a virtual vacuum. A massive explosion on the Moon could not be "heard" as such on Earth simply because there would be no medium to transmit the sound waves. On the other hand, energy from the explosion may be transmitted in the form of ejected particles, and on reaching the Earth's surface could be translated by an appropriate receiver into a sound-wave pattern. It would be unwise, therefore, to dismiss out of hand the claims of those who say they have "heard" the aurora, despite the fact that there is sensibly no atmosphere to convey the sound waves to the ground.

When is the Best Time to See the Aurora?

I am frequently asked this question. In astronomy one can divide subjects between the predictable and the unpredictable. For example, tables are issued many years in advance giving the positions for planets and other objects in the Solar System to a high degree of precision. Also within the category of predictable events are solar and lunar eclipses and occultations of stars by the Moon and other bodies in the Solar System.

Amongst the non-predictable events are the appearance of the aurora itself; sunspots; visitations by comets hitherto unknown; transient phenomena on other planets; and the sudden appearance of bright "new" stars (novae and supernovae). However, as far as the aurora are concerned we are not completely in the dark (pun intended) as will be understood from an analysis of auroral events already mentioned.

So my advice to would-be observers is to look out at the sky whenever you have the opportunity. Of course twilight is too strong during mid-summer in Orkney for aurora to show even if they were active. However, by mid-August conditions improve from this point of view. Constant vigilance is the key for the practical astronomer. I have seen auroral forms on nights when neither the magnetometer nor evidence from solar observations would indicate an imminent display.

Another frequently asked question relates to the appearance of aurora at lower latitudes. Again nothing can be ruled out. Reports of the great aurora of September 1859 were received from ships that were located well away from the Earth's polar

This image was taken on the evening of 8 October 2015 and shows a bright auroral 'patch' obscuring part of the Milky Way. Two prominent 'corners' of the famous Summer Triangle – the bright stars Vega (top right) and Altair (bottom left) – are clearly seen, the third corner (Deneb) being excluded from the frame. (John Vetterlein)

Some intense auroral displays were observed from Orkney during and around the time of the 2014 solar maximum. This view captures the aurora close to the zenith on the night of 27/28 February. During this particular display, dark red and lighter shades were seen rising to elevations of 70° in the west and northwest. The distinctive constellation Auriga, together with its leading star Capella, can be seen at top left of image. (John Vetterlein)

regions. Even during less violent displays, observations of aurora have come from as far south as northern Spain and other locations at similar latitudes.

Cautionary Note

In discussing the close linking of auroral activity with the Sun a note of caution has to be introduced regarding the Sun itself. Never observe the Sun with or without optical aid unless this is via the projection method or with the use of proper and correctly-fitted filters. To look at the Sun without suitable protection can lead to permanent blinding due to the fact that optical devices concentrate the Sun's light and heat into a small area, thereby raising the temperature to inflammable values within seconds.

How to Read a Scientific Paper

David Harper

Professional astronomers publicise their work in many different ways. They appear on television, give talks to local astronomy societies, and write articles for popular astronomy magazines and for books such as the *Yearbook of Astronomy*. When communicating with other professional astronomers, they may give talks at conferences, or seminars when they visit colleagues at other universities, but the most important medium for publishing new research is the scientific journal.

In this article, I will explain why scientific journals are so significant to professional astronomers, how a scientific paper is written and published, and why it is easier than ever before for amateur astronomers to access this valuable source of information.

What are Journals?

Journals are the collective memory of the scientific community. Their purpose is to publish new discoveries and other research. They establish priority: who made a particular discovery, and when. They record what research has already been done, to avoid duplication. Moreover, before air travel and the Internet made the sharing of news and information so easy, journals were the fastest way to announce new research.

There is also a more pragmatic reason why scientists publish their research in journals. Scientific research is an expensive business. Whilst amateur astronomers can make valuable contributions, much astronomical research requires access to very large telescopes or satellites or space probes to make the basic observations. Astronomers also need powerful computer

The Royal Astronomical Society publishes *Monthly Notices*, the oldest English-language journal devoted to astronomy. (Mike Peel (www.mikepeel.net)/ CC-BY-SA-4.0 licence)

systems to analyse their observations. The money which pays for most astronomical research comes from governments: in the U.K., the Science and Technology Facilities Council (STFC) is the government agency which awards research grants to astronomers working in British universities and pays subscriptions to multi-national partnerships such as the European Southern Observatory, allowing British astronomers access to facilities such as the Very Large Telescope in Chile. Funding agencies such as the STFC require evidence that their research grants are being used effectively, and this is measured by the number of papers that are published in scientific journals as the result of each grant.

In many countries, including the U.K., lecturers and professors in universities are evaluated primarily on the research that they carry out, and this is measured by the number of papers that each individual publishes. Jobs can be lost if too few papers are published, and promotion depends on a strong publication record.

Which Journals are Important in Astronomy?

The oldest English-language scientific journal in the world is *Philosophical Transactions of the Royal Society*. Published by the Royal Society, this first appeared in 1665 and is still in print today. It contains papers on a wide range of subjects, including astronomy. For example, James Bradley's announcement of his discovery of the aberration of starlight by the motion of the Earth around the Sun was announced in *Philosophical Transactions* in 1728.

Today, most astronomical research is published in a small number of journals devoted specifically to astronomy:

Monthly Notices of the Royal Astronomical Society was first published in 1827, making it the oldest English-language astronomy journal. Also known by its abbreviation, MNRAS, it publishes papers in all areas of astronomy. Despite the title, you do not have to be a Fellow of the Royal Astronomical Society (RAS) to publish a paper in *Monthly Notices*.

James Bradley (1692–1762) was the third Astronomer Royal, and a Fellow of the Royal Society. He announced his discovery of the aberration of starlight in the Society's journal *Philosophical Transactions*. (John McCue)

- *The Astronomical Journal* was founded by the American astronomer Benjamin Apthorp Gould in 1849. It is published on behalf of the American Astronomical Society, and is known colloquially as AJ.
- *The Astrophysical Journal* is also a journal of the American Astronomical Society. It was first published in 1899. To distinguish it from its older sibling, it is known as ApJ.
- *Astronomy and Astrophysics* was created in 1969 by the merger of several European journals, including the French *Annales d'Astrophysique* and the German *Zeitschrift für Astrophysik*. It is published on behalf of the European Southern Observatory organisation, and is known by the abbreviation A&A. Until 1999, there was a companion journal, *Astronomy and Astrophysics Supplement Series*, where papers reporting observational research or describing new methods of observation or computing were published. These papers now appear in the main journal.

The American astronomer Benjamin Apthorp Gould (1824–1896) founded *The Astronomical Journal* in 1849. (Harper's Encyclopædia of United States History, Vol. IV (1905)/ Harper & Brothers)

There are also journals devoted to specific areas of astronomy. A prominent example is *Icarus*, which publishes papers on planetary science.

Annual Review of Astronomy and Astrophysics has been published since 1963. It contains longer articles which summarise the current state of knowledge about a particular area of astronomy. Each article is written by astronomers with many years of experience in the subject that it covers, and includes extensive references to the most important research papers in that subject. There are also autobiographical essays by distinguished professional astronomers, looking back over their careers.

If a discovery is sufficiently important, it may be published in *Nature* or *Science*. These are general science journals which only accept papers containing research of the highest importance. A paper in either of these journals is a career-enhancing achievement for any professional scientist.

There are also many journals which are produced by non-professional astronomers. The leading example in the U.K. is the *Journal of the British Astronomical Association*, which is published by the BAA to share observations and discoveries

The headquarters of the European Southern Observatory (ESO) in Garching bei München, Germany. ESO publishes the journal *Astronomy and Astrophysics*. (E.Graf (graf-flugplatz.de) / ESO)

made by its members. Another is *The Antiquarian Astronomer*, the journal of the Society for the History of Astronomy. Both of these journals employ the peer-review process to maintain the highest standards of quality and originality in the articles they publish.

In addition, the American Association of Variable Star Observers, the Association of Lunar and Planetary Observers and the International Occultation Timing Association each publish their own journal series to record and share the work of their members.

How Do Papers Get Published?

All of the professional journals mentioned above use the peer-review system. This means that every paper submitted to the journal is reviewed by experts in the subject area to ensure that the research which it describes is original, of sufficient importance, has been done correctly, and is being reported clearly. For reputable journals, the peer-review process is a vital filter which ensures that only original, high-quality research is published.

Practice varies between journals. Some editors rely on a single reviewer for each paper, while others send the paper to two or three. Some journals allow the reviewers to remain anonymous, whilst others do not. Papers may be rejected outright, if they repeat research that has already been done before or if the research is of poor quality. More often, the reviewers will recommend changes to the paper before it can be published. They may even insist that the authors repeat some aspects of the research using a different approach, and then submit a revised version of the paper.

Most journals also impose a strict limit to the length of a paper, often ten pages or less. This encourages authors to describe their research as succinctly as possible. Before the advent of electronic publishing, journals were sent to university libraries in printed format, and a strict length limit helped to control the cost of production and distribution. Reviewers may still advise that a paper is too long, and recommend that it be shortened before it can be accepted for publication.

The Anatomy of a Scientific Paper

A typical scientific paper has a well-defined structure.

- **Title**. The title of the paper describes its purpose in a single sentence.
- **Authors**. The authors of the paper are listed, along with the addresses of the universities or observatories where they work. Their email addresses may also be given. The names of the authors are usually listed in an order which signifies their relative contribution to the work: readers will assume that the first-named author made the most significant contribution, so to be the first author of a paper confers special status.
- **Received and Accepted dates**. The date when the first draft of the paper was received by the journal is shown, as well as the date when final revised version was accepted for publication. These dates are important in establishing priority if two groups of researchers publish similar findings.
- **Abstract**. Every paper has an abstract. This summarises the paper in one or two paragraphs, allowing the reader to decide whether or not the paper is relevant to them.
- **Introduction**. This places the paper in context. It references previous research and explains how the work described in the current paper builds on, improves or overturns, that research. It may also explain why the current research was undertaken, but it will not mention any results or conclusions.
- **Methods**. This section describes how the research was carried out. If the paper is reporting observational work, then the methods section must explain how

the observations were made. If it is reporting theoretical work or an analysis of existing observations, it should show the mathematical derivation of the theory or the methods used in the analysis. In all cases, the description must be detailed enough to allow other researchers to repeat the work themselves, given similar facilities.

- **Results**. This section lists the observations that were made, or summarises the significant outcomes of a piece of theoretical research, or reports improved values of parameters obtained from an analysis of existing observations. These can range from more accurate orbital elements for a planetary satellite to an improved value of the age of the universe. In all cases, they must include estimates of the uncertainty of the result, calculated using standard statistical methods. They should be compared with the results which have been obtained by other researchers in the recent past.

- **Conclusions**. At the end of the paper, the authors will summarise the results that they have reported. They may note that their results agree with previous work, but at a higher level of accuracy or confidence. Alternatively, they might highlight significant disagreement with results obtained by other researchers, and recommend that further research needs to be carried out to resolve the discrepancy. This is not unusual, nor is it grounds for a paper to be rejected. After all, major scientific advances sometimes result from such unexpected discrepancies.

- **Acknowledgements**. It is customary to thank the funding agency or agencies which provided the money to allow the research to be carried out. Other individuals or organisations that provided help or facilities may also be thanked, as may the reviewers.

- **References**. Throughout the body of the paper, the authors will have referred to previous work carried out by other researchers. They may also have referred to their own earlier work. The reference section of the paper, which is always at the end, lists all of the sources that have been referenced. These will normally be other peer-reviewed papers, although books and other sources may sometimes appear. The list will be in a standard bibliographic format, which I describe below. The electronic version of a paper will normally provide links which take the reader directly to the electronic versions of the papers in the list.

Decoding a Reference to a Paper

This is a typical entry that might appear in the reference section of a paper:

*Zhang, H.Y., Shen, K.X., Dourneau, G., Harper, D., Qiao, R.C., Xi, X.J., Cheng, X., Yan, D., Li, S.N., Wang, S.H., (2014) MNRAS **438**, 1663*

The authors are listed first, exactly as they appear at the top of the paper. In this case, Zhang and Shen are the first-named authors, so we may infer that they made a greater contribution to the research than Dourneau or Harper, who in turn contributed more than the other six co-authors. The year of publication follows the list of authors. When this paper is cited by other researchers, they will refer to it as *Zhang et al (2014)* in the body of their own paper, and include the full reference only at the end of their paper.

Finally, there is the name of the journal (often abbreviated), the volume, and the page or article number. This paper can be found in *Monthly Notices of the RAS*, volume 438, starting at page 1663. As journals have moved from printed volumes to online publishing, page numbers are being replaced by article numbers. *Astronomy & Astrophysics* made this transition in 2010, with *The Astronomical Journal* and *The Astrophysical Journal* following in 2011, but *Monthly Notices of the RAS* still uses page numbers.

The title of the paper is rarely included. This paper's title is *An orbital determination of Triton with the use of a revised pole model*, but you would need to find the paper at the *Monthly Notices* web site to discover that.

Closed Access versus Open Access

Before the advent of electronic publishing, it was difficult for anyone outside a university or research institute to obtain copies of scientific papers. Today, it is very much easier. All of the major journals have web sites where they make the content of both current and past issues available online.

Most journals are published by commercial companies which charge for access in much the same way as major newspapers such as the *New York Times*. University libraries pay subscriptions to a wide range of journals, allowing members of the university to access them freely. Fellows of the RAS have full access to *Monthly Notices* as a benefit of membership, and technical members of the Astronomical Society of the Pacific likewise have full access to the Society's premier research journal, *Publications of the Astronomical Society of the Pacific*.

In January 2022, the American Astronomical Society made the full contents of both of its flagship journals, *The Astronomical Journal* and *The Astrophysical Journal*, freely available to anyone, a practice known as open access. In March 2023, the RAS announced that Monthly Notices would become open-access from 2024, but authors would have to pay a processing fee of £2,310 per paper from that date. A number of newer journals were founded with an open access policy, including the Public Library of Science (PLoS) family. Other publishers often make papers freely available once a period of time has elapsed. The authors of papers can also pay an extra fee to the publisher so that the paper is available to anyone as soon as it is published. Even if a paper is not freely available, a polite email to the leading author will often result in a PDF copy of the paper.

Where Can I Find Interesting Papers?

Most Wikipedia articles on astronomical topics will include citations of relevant research papers, together with direct links to online versions of the papers in the reference section. There may be several links to each paper:

- **arXiv.org** is a web site where you can find preprints of papers. A preprint is an early version of a paper, which may not have been submitted to a journal yet. Many physicists and astronomers send preprints to arXiv so that they can receive informal feedback from the scientific community before submitting a more polished version to a journal. Access to preprints at arXiv is free.
- Bibcode links to the NASA Astrophysics Data System online library of papers. It provides free access to the full text of several million astronomy papers, including many which are of great historical interest.
- DOI is short for "digital object identifier". This is an international standard for assigning unique and permanent identifiers to online resources such as journal papers. Most major journals assign a DOI to each paper when it is published.

Monthly Notices
of the
ROYAL ASTRONOMICAL SOCIETY

MNRAS **438**, 1663–1668 (2014)   doi:10.1093/mnras/stt2304
Advance Access publication 2013 December 21

An orbital determination of Triton with the use of a revised pole model

H. Y. Zhang,[1,2]★ K. X. Shen,[1] G. Dourneau,[3,4] D. Harper,[5] R. C. Qiao,[1] X. J. Xi,[1,2] X. Cheng,[1,2] D. Yan,[1,2] S. N. Li[6] and S. H. Wang[6]

[1]*National Time Service Center (NTSC), Chinese Academy of Sciences, PO Box 18, Lintong, Shaanxi 710600, China*
[2]*Graduate University of the Chinese Academy of Sciences, Beijing 100039, China*
[3]*Univ. Bordeaux, LAB, UMR 5804, F-33270 Floirac, France*
[4]*CNRS, LAB, UMR 5804, F-33270 Floirac, France*
[5]*Wellcome Trust Sanger Institute, Wellcome Trust Genome Campus, Hinxton, Cambridge CB10 1SA, UK*
[6]*Shanghai Astronomical Observatory (SHAO), Chinese Academy of Sciences, Shanghai 200030, China*

Accepted 2013 November 28. Received 2013 November 24; in original form 2013 July 2

ABSTRACT
In this paper, we used the 3108 Earth-based astrometric observations from the Natural Satellite Data Center over more than 30 yr time span from 1975 to 2006 for determining the epoch state vectors of the Neptunian largest satellite Triton. These observations almost contain all modern photo and CCD observations available. In integrating perturbation equation, the barycentric frame of Neptune–Triton system is adopted, and in considering the oblateness perturbation due to Neptune, a revised pole model describing the precession of the Neptune's pole is used in our calculation.

Key words: astrometry – planets and satellites: general.

All papers begin with the title, author list and a short abstract. This is the *Monthly Notices* paper mentioned in the article. (Royal Astronomical Society / *Monthly Notices*)

A typical DOI looks like this: **10.1093/mnras/stt2304** and is normally given as part of a web address e.g. **https://doi.org/10.1093/mnras/stt2304**. This will always re-direct your browser to the current location of the paper at the publisher's web site.

The NASA Astrophysics Data System allows the user to search for papers about astronomy and astrophysics by any combination of year, author, key word, astronomical object, or words in the abstract. It is a very convenient way to find papers about a specific astronomical object or by a particular astronomer. Its catalogue includes papers in *Nature* and *Science* as well as all of the leading astronomy journals, both past and present.

Making Sense of It All

Scientific papers can often seem impenetrable, even to a reader who is an active researcher in the same branch of science. Most papers are written by researchers who have spent many years becoming an expert in a very narrow and specialised field, and they are writing for other experts in that field who share the same deep background knowledge. Each area of astronomy has its own jargon, so a paper describing the latest work in theoretical cosmology may be as baffling to a professional planetary scientist as it is to most amateur astronomers. Astronomy is also a highly mathematical subject at the professional level, and many papers contain mathematics which is at advanced A-level standard or above, requiring an understanding of calculus and statistics. Having said that, it can be very rewarding to read a scientific paper, even if you don't understand every nuance and technical detail. It offers a fascinating insight into the way that research is done by professional scientists. The Internet provides many resources which can help to clarify technical jargon and to explain topics such as stellar evolution or orbital mechanics in plain language.

Further Reading

There are many textbooks aimed at students in the first year of an astronomy degree which are very readable. *An Introduction to Modern Astrophysics* by Bradley Carroll and Dale Ostlie or *An Introduction to Astronomy and Cosmology* by Ian Morison both provide a good foundation for anyone wishing to read papers on current research in most areas of astronomy.

Small Stars

John McCue

Making the best of a bad job was clearly the most unexpected skill that English physicist James Chadwick, discoverer of the neutron, possessed. Born in 1891 in Bollington, Cheshire, he studied under Ernest Rutherford at Manchester University – from where he graduated in 1911 – before setting off for Berlin in 1913 to work with the German physicist Hans Geiger. History recalls that World War One erupted one year later leaving Chadwick imprisoned as an enemy alien. He passed those four war years living and working in a stable, but, undeterred and unbeaten, he kept his research skills up to scratch with investigations into oxidation and photochemistry.

James Chadwick, discoverer of the neutron. (John McCue)

With the Great War at a close, he returned to Manchester to carry on working with Ernest Rutherford, who in 1917, while Chadwick was locked up, sensationally found that the atom was largely empty space, an incredibly small nucleus made up of positively charged protons, surrounded by a cloud of negatively charged electrons. He used the radioactivity known as alpha particles which had a positive charge and sometimes bounced off the nucleus of target atoms at crazy angles leading Rutherford to that conclusion. The two scientists continued in this vein, managing to split some of the lighter atoms with these alpha particles. From 1920, with astute foresight, Rutherford had suspected that there was another particle in the nucleus. He was right, and Chadwick found it in 1932.

Setting up beryllium as a target for his alpha particle bullets, Chadwick saw radiation coming from the collisions which was much more energetic than he expected. Replacing beryllium in the target with boron, for more information, he eventually realised it was not radiation but a beam of particles, as Rutherford had thought, and they were neutral particles. The neutron was slightly more massive than the proton, rightly so, as the neutron had to carry an electron within it to neutralise the positive charge. At that stage, the nuclear structure of all atoms heavier than simple hydrogen was settled as a mixture of protons and neutrons, mostly in equal numbers. Years later, the protons and neutrons would reveal a sprite-like inner structure of quarks, but that takes us beyond our story…

The humble, uncharged neutron finally stole the limelight as the stuff of the smallest stars known – neutron stars. These stars begin life massive, more than eleven times that of our sun, and end their existence in supernova explosions which, incidentally, enrich the surrounding interstellar clouds with the heavy elements produced inside these stars, so providing the raw materials for the next generation of stars, their planets, and any life forms that may then emerge. Even though huge, the early life of a massive star proceeds as any other *sun-like star*, converting hydrogen to helium to pour that fusion energy out into the space around it, nurturing any planets it may have. When the hydrogen fuel eventually runs out the helium – Lego-like – constructs more and more complicated and heavier elements.

Eventually, by virtue of its size, the massive star builds iron. This well-known element is the most stable of all atomic nuclei, and it resists fusion with elements of any kind – so no more outflow of energy! Without fusion energy pushing outwards to balance gravity pulling inwards – stellar arm-wrestling – the core of the star gives in and begins to contract. To add a crisis to this drama, the energy of this gravitational collapse causes a rise in temperature, pressure, and density until the electrons are stripped from their nuclei. This shrinkage would normally be halted by quantum mechanics which forbids the electrons being squashed any more (electron degeneracy) but the shrinking goes on because the star's too big. It, or to be precise, its core, must be less than 1.4 times the mass of the sun in order to evade collapse, a figure worked out by the Indian-American astrophysicist Subrahmanyan Chandrasekhar in the 1930s when he was only in his 20s. The core would then be stable, a white dwarf, about the size of the earth, and a football made of its material would weigh about the same as a thousand elephants on earth. But our massive star carries on dwindling, regardless …

Gravity, unopposed, now squeezes the star relentlessly, overcoming the electron pressure. How will this collapsing gravitational energy be released? Something must give. Firstly, gamma rays from the incredibly hot collapse can punch their energy into an iron nucleus and break it into thirteen helium nuclei (which in turn break up into protons and neutrons as the temperature increases further) and four neutrons. Secondly, the core's dramatic crumple gives some electrons so much energy that they can penetrate the iron nuclei and combine with the protons therein, crucially forming neutrons. With each neutron created, a neutrino is released, which escapes the star, carrying away its energy. This escape of energy concedes even more to gravity, and so accelerates the rapid crush of the core.

Finally, the density becomes so high – neutron degeneracy – that its gravity can't overcome it, and the neutron star is born. In the energy count, the fireworks of the subsequent supernova blast is a diminutive side-show compared to this enormous release of gravitational energy which heralds the small star's birth.[1] Remember the thousand-elephant football sample of white dwarf material? The football portion of a neutron star clocks in at about a hundred billion elephants!!

The upper limit for the mass of a neutron star is three times the mass of the sun. Any more than that and gravity wins over every possible obstacle, and a black hole is the result. A new and potentially game-changing observation has been made recently by Victor Doroshenko with the space observatories *XMM-Newton* and *Suzaku*. The well-known supernova remnant HESS J1731-347 in the constellation Scorpius has, of course, a neutron star at its centre, and six years of X-ray observations have led to the conclusion that its mass is about 0.8 times that of the Sun.[2] The gravity of such a dwarf star could never overcome electron degeneracy, and it should be a white dwarf. The jury is out, and a verdict hangs in the balance …

When the collapse suddenly hits this neutron brick wall though, it rebounds outwards in the aforementioned shock wave explosion, a supernova, during which time all the elements beyond iron are manufactured and scattered to interstellar space. The core itself – the neutron star – now spins faster to conserve angular momentum as it shrinks, in just the same way that an ice-skater spins faster when arms are pulled in. The final size, about 15 kilometres in diameter, the size of a city, is now miniscule compared to the white dwarf, and the final rotation is

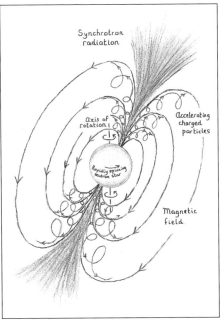

The rapidly spinning pulsar shows flashes of radiation towards the earth from its magnetic poles. (John McCue)

1. *The Physics of Stars, 2nd. Edition*, Phillips, A.C., 1999, John Wiley & Sons.

2. 'A strangely light neutron star within a supernova remnant', Victor Doroshenko, reported by Monica Young, *Sky and Telescope*, March 2023, page 9.

incredibly fast – young neutron stars can rotate hundreds of times a second. As a sub-species of neutron star, these spinners are called pulsars.

Every star possesses some magnetic field. When this field is now concentrated over a smaller surface area after the star collapses, the magnetic field is correspondingly stronger. Incidentally, if the earth were crushed down to neutron star density, it would be about the size of a football ground. This strong, spinning magnetic field is responsible for the acceleration of charged particles outward from the neutron star's magnetic poles. Such acceleration causes the emission of synchrotron radiation from the particles, in the same way that a radio wave is generated here on earth if electrons are pushed up and down an aerial. We pick up these pulses, like a flashing lighthouse, across the depths of space as the neutron star furiously spins, hence its name – pulsar. Clearly, the radiation pulses from the spinning neutron star will only be seen if the magnetic poles are angled to some extent towards our line of sight from earth.

Jocelyn Bell Burnell, discoverer of the pulsar species of neutron star. (John McCue)

The young Jocelyn Bell Burnell, at the tender age of 24, was the first to find these flashes from a neutron star. Her father was an architect with the Armagh Observatory and the staff there encouraged the child with her early interest in astronomy. Born in Belfast in 1943, she graduated from the University of Glasgow in 1965 before heading to Cambridge for her Ph.D. research under the supervision of Anthony Hewish. These historic years saw her ground-breaking find. First though, she actually had to build the radio telescope! Made up of fixed antennae, it covered four-and-a-half acres, equivalent to a radio dish with a diameter of 150 metres, was extremely sensitive, and recorded fast variations in any signals it received. It was designed in this way for the study of quasars. The sky survey began in 1967.

One morning, looking at pen recorder signals from the previous night's radio observations, she saw rapid spikes. She thought it might be local radio interference, but she realised that the source had a fixed position in the sky, so could not be terrestrial. The pulses occurred every 1.337 seconds and were as regular as the most accurate clock on Earth. These were pulsars.

The fastest pulsar known so far is designated PSR J1748-2446ad. Over 20,000 light years away in the constellation of Sagittarius, it spins at 716 times per second

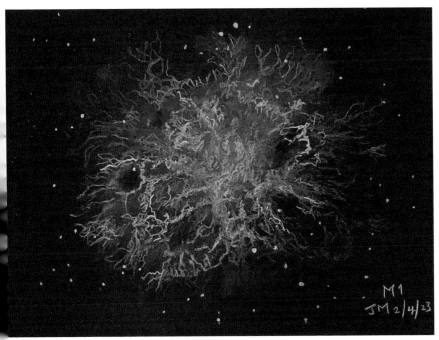

An artistic impression of Messier 1, the Crab Nebula, at the centre of which lies the most famous pulsar of all. (John McCue)

(twice as fast as a smoothie blender), and was discovered by Jason Hessels in 2004[3] using the giant 100-metre Green Bank radio telescope. The most famous pulsar is at the centre of the Crab Nebula, which exploded as a supernova and was observed by Chinese astronomers in 1054 CE. Its spin rate is about 30 times a second (increasing by a millisecond every 90 years) so it's much slower, and hence older, than PSR J1748-2446ad.

At first Bell Burnell and Hewish thought her discovery might be a regular signal from an extra-terrestrial civilisation, a fascinating possibility in itself, but she found several more pulsars with similar spike rates, but their periods varying very slightly. Now over 3,000 pulsars are known, with periods varying from milliseconds to a few seconds.

A pulsar's period can gradually change in one of two ways. It can increase as the synchrotron radiation from the magnetic poles of the neutron star gradually

3. 'A Radio Pulsar Spinning at 716 Hz', Hessels, J.W.T., et al., 2004, *Science*, **311**, 5769, 1901–1904. Article available at: **arxiv.org/pdf/astro-ph/0601337.pdf**

dissipates the star's rotational energy, slowing it down. The period can gradually decrease if the pulsar is in a mutual orbit with a companion star. The pulsar's immense gravitational pull on the unfortunate partner star will rip material from it, which spirals inwards and onto the pulsar, speeding it up.

Another sub-species of neutron star is the magnetar, the magnetic field of which is about a thousand times that of a 'normal' pulsar. These stars were predicted in 1992 by American astronomers Robert Duncan and Christopher Thompson.[4] Such magnetism would pull a set of keys out of your pocket at a distance of 200,000 kilometres or halfway to the moon. There are theories explaining this phenomenal field strength. Strong convection currents of charged particles (some of which are still there as well as the neutrons) will be set up in the superfluid interior of the forming neutron star and they will generate a magnetic field in the same way that our earth does. If there is a match between the rise and fall of the convection currents and the spin of the pulsar, like pushing a child on a swing, then the overall magnetic field is amplified, even by about a thousand times. On the other hand, a magnetar could merely be the collapse of a star that originally had an unusually high magnetic field.[5]

Neutron stars are as dense as any physical object in the universe could possibly be, and for mere humans to comprehend that density might seem impossible. However, if you consider how strong the massive neutron star's gravity must be to stop it flying apart under such prodigious spinning, you're part of the way there.

4. 'Formation of Very Strongly Magnetized Neutron Stars: Implication for Gamma-Ray Bursts', Duncan, R. C. and Thompson, C., 1992, *The Astrophysical Journal Letters*, **392** L10–L13. Article available at **adsabs.harvard.edu/full/1992ApJ...392L...9D**

5. 'Magnetars', McCue, J., *Yearbook of Astronomy 2023*, 121–124, White Owl.

Miscellaneous

Some Interesting Variable Stars

Tracie Heywood

You may have considered taking up variable star observing but how should you choose which stars to observe? There are so many variable stars in the night sky and you don't want to waste your time attempting to follow the "boring" ones. Your choice of stars will, of course, depend on the equipment that you have available, but also needs to be influenced by how much time you can set aside for observing.

This article splits some of the more interesting variable stars into three groups. The group that is most suited to you will depend on how often you can observe each month and for how long you can observe on a clear night. The light curves included have been constructed from observations stored in the Photometry Database of the British Astronomical Association Variable Star Section. Comparison charts for most of these stars can be found on the BAA Variable Star Section website at **britastro.org/vss**

One-Nighters

These are stars that can go through most of their brightness variations in the course of a single (reasonably long) night and would suit *people who can only observe occasionally, but can then observe well into the night.*

STAR	TYPE	RA		DEC		MAX / MIN	PERIOD
		H	M	°	′		
RZ Cassiopeiae	EA	02	49	+69	38	6.4 / 7.8	1.1952503 days (~29 hours)
VW Cephei	EW	20	37	+75	36	7.3 / 7.8	0.2783089 days (~6.68 h)
Beta (β) Persei (Algol)	EA	03	08	+40	57	2.1 / 3.4	2.867 days (~69 hours)
RW Tauri	EA	04	04	+28	07	8.0 / 11.6	2.7687937 days

RZ Cassiopeiae is an Algol-type eclipsing variable. Primary eclipses last for just under 5 hours and are 'V-shaped', indicating that they are partial eclipses. Predictions for upcoming eclipses can be found at **www.as.up.krakow.pl/minicalc/CASRZ.HTM**

VW Cephei is a W UMa type eclipsing variable. These are 'contact' binary systems that show no period of constant brightness between eclipses. Predictions for upcoming eclipses can be found at **www.as.up.krakow.pl/minicalc/CEPVW.HTM**

Beta Persei (Algol) is an eclipsing variable that shows deep primary eclipses but only very shallow secondary eclipses. Although circumpolar from the UK, it is too low in the sky for observation from April to June.

RW Tauri is another Algol-type eclipsing variable star. Predictions for upcoming eclipses can be found at **www.as.up.krakow.pl/minicalc/TAURW.HTM**

Most Clear Nights

These are stars that vary a bit more slowly, but which can display significant changes over a week or two. They would suit *people who can observe for a short while on (nearly) every clear night.*

STAR	TYPE	RA		DEC		MAX / MIN	PERIOD
		H	M	°	'		
OJ+287 Cnc	BL Lac	08	55	+20	07	12.7 / 17.0	None
CH Cygni	Z And	19	25	+50	15	5.4 / 10.5	97 days, ~4700 days
AM Herculis	Polar	18	16	+49	52	12.3 / 15.7	3.094 hours (orbital)
Beta Lyrae	EB	18	50	+33	22	3.4 / 4.3	12.944 days

OJ+287 is a BL Lac object, an active galactic nucleus that is believed to contain a binary black hole. Located around five billion light years from Earth, it produces optical outbursts from time to time.

CH Cygni is a binary system in which one star shows semi-regular periodic variations, while the other star is an eruptive variable star that shows erratic changes.

AM Herculis is a Polar, a cataclysmic binary system in which the white dwarf's magnetic field dominates the flow of matter between the two stars. It shows a mix of "high" (brighter) and "low" (fainter) states.

Beta Lyrae is an eclipsing variable. Both primary and secondary eclipses can be monitored visually and there are also small changes in brightness between eclipses.

Several Times per Month

Slower variables whose brightness will change significantly over several months or a year. These variables would suit *people who can observe several times per month, but not necessarily on every clear night.*

STAR	TYPE	RA		DEC		MAX / MIN	PERIOD
		H	M	°	′		
R Cassiopeiae	Mira	23	58	+51	23	4.8 / 13.5	430 days
RW Cephei	Semi Reg	22	23	+55	58	6.2 / 7.8	346 days
T Cephei	Mira	21	10	+68	29	5.4 / 11.0	389 days
V973 Cygni	Semi Reg	19	45	+40	43	6.1 / 6.8	40 days, 237 days
RS Coronae Borealis	Semi Reg	15	59	+36	01	7.3 / 9.5	332 days
Alpha Herculis	Semi Reg	17	15	+14	23	2.9 / 3.7	93 days, 126 days
Z Ursae Majoris	Semi Reg	11	57	+57	52	6.2 / 9.9	185 days, 196 days

R Cassiopeiae is a Mira type variable. As for other Mira type variables, some maxima are brighter than others. The 2023 maximum peaked at magnitude 5.6, while that of 2015 only reached magnitude 8.3. The 2025 maximum is predicted to occur during July.

T Cephei is another Mira type variable, lying not far from Beta Cephei. Maximum in 2025 is predicted to occur during early July.

RW Cephei, **V973 Cygni**, **RS Coronae Borealis**, **Alpha Herculis** and **Z Ursae Majoris** are semi-regular variables that can be followed using binoculars. "Semi-regular" indicates that the brightness variations repeat only roughly from one cycle to the next. The observed brightness range differs between cycles, sometimes covering the whole of the listed range (occasionally exceeding it), but more often being somewhat smaller.

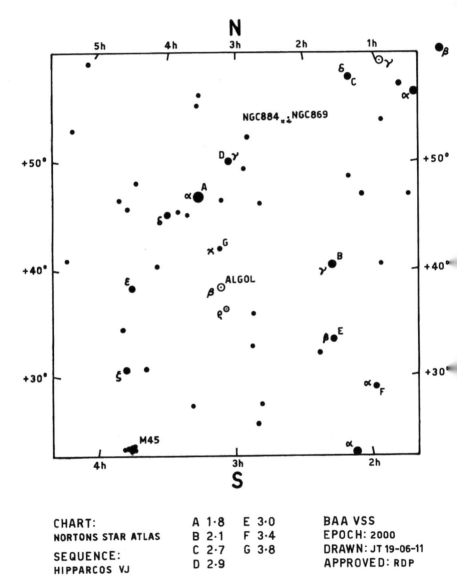

The BAA VSS finder chart for Beta (β) Persei (Algol). (BAA Variable Star Section)

Minima of Algol in 2025

Beta (β) Persei (Algol): Magnitude 2.1 to 3.4 / Duration 9.6 hours

	Day	h	*		Day	h	*		Day	h	*		Day	h	*
Jan	2	7.5		Feb	2	20.5	*	Mar	3	12.7		Apr	1	4.8	
	5	4.3			5	17.3			6	9.5			4	1.6	
	8	1.1	*		8	14.1			9	6.3			6	22.4	*
	10	22.0	*		11	10.9			12	3.1			9	19.3	
	13	18.8	*		14	7.8			14	23.9	*		12	16.1	
	16	15.6			17	4.6			17	20.7	*		15	12.9	
	19	12.4			20	1.4	*		20	17.6			18	9.7	
	22	9.2			22	22.2	*		23	14.4			21	6.5	
	25	6.0			25	19.0	*		26	11.2			24	3.3	
	28	2.9			28	15.8			29	8.0			27	0.2	
	30	23.7	*										29	21.0	

	Day	h	*		Day	h	*		Day	h	*		Day	h	*
May	2	17.8		Jun	3	6.8		Jul	1	22.9		Aug	2	11.9	
	5	14.6			6	3.6			4	19.7			5	8.7	
	8	11.4			9	0.4			7	16.6			8	5.5	
	11	8.2			11	21.2			10	13.4			11	2.4	*
	14	5.1			14	18.0			13	10.2			13	23.2	*
	17	1.9			17	14.9			16	7.0			16	20.0	
	19	22.7			20	11.7			19	3.8			19	16.8	
	22	19.5			23	8.5			22	0.6	*		22	13.6	
	25	16.3			26	5.3			24	21.5			25	10.4	
	28	13.1			29	2.1			27	18.3			28	7.3	
	31	10.0							30	15.1			31	4.1	

	Day	h	*		Day	h	*		Day	h	*		Day	h	*
Sep	3	0.9	*	Oct	1	17.0		Nov	2	6.0		Dec	3	19.0	*
	5	21.7	*		4	13.9			5	2.8	*		6	15.8	
	8	18.5			7	10.7			7	23.7	*		9	12.6	
	11	15.3			10	7.5			10	20.5			12	9.5	
	14	12.2			13	4.3	*		13	17.3			15	6.3	
	17	9.0			16	1.1	*		16	14.1			18	3.1	*
	20	5.8			18	21.9	*		19	10.9			20	23.9	*
	23	2.6	*		21	18.8			22	7.7			23	20.7	*
	25	23.4	*		24	15.6			25	4.6			26	17.5	*
	28	20.2	*		27	12.4			28	1.4	*		29	14.3	
					30	9.2			30	22.2	*				

...ses marked with an asterisk (*) are favourable from the British Isles, taking into account the altitude of the ...ble and the distance of the Sun below the horizon (based on longitude 0° and latitude 52°N).

...mes given in the above table are expressed in UT/GMT.

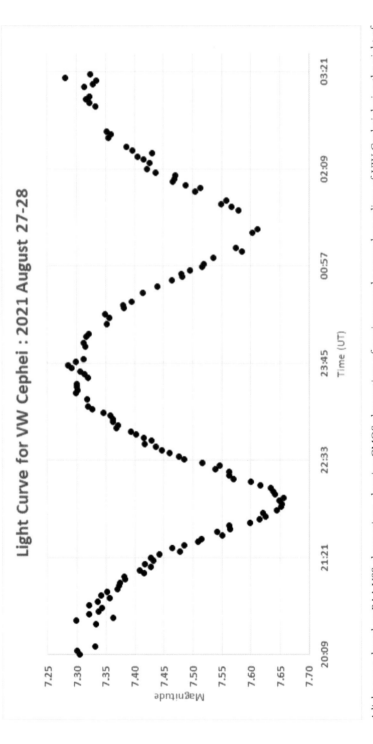

A light curve, based on BAA VSS observations, showing CMOS observations of a primary and a secondary eclipse of VW Cephei during the night of 27/28 August 2021. (BAA Variable Star Section)

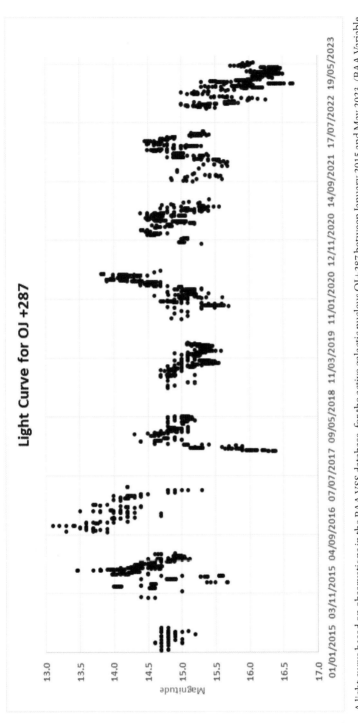

A light curve based on observations in the BAA VSS database, for the active galactic nucleus OJ+287 between January 2015 and May 2023. (BAA Variable Star Section)

Light Curve for AM HER

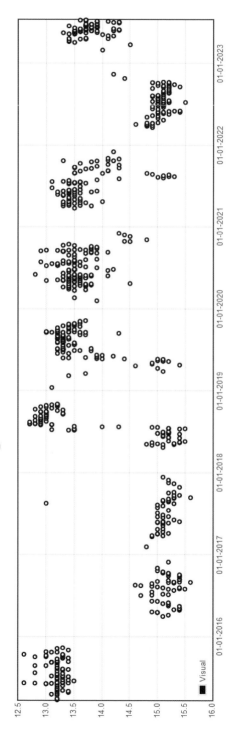

The BAA VSS light curve for the cataclysmic variable AM Herculis, showing visual observations made between January 2015 and June 2023. (BAA Variable Star Section)

Light Curve for RW CEP

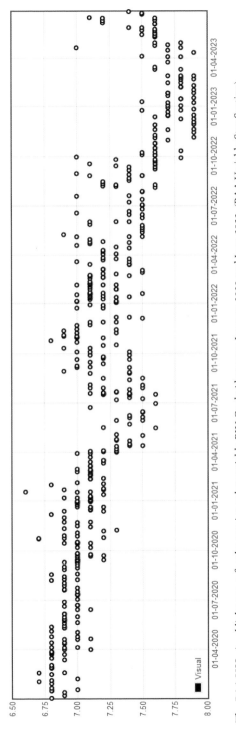

The BAA VSS visual light curve for the semi-regular variable RW Cephei between January 2020 and June 2023. (BAA Variable Star Section)

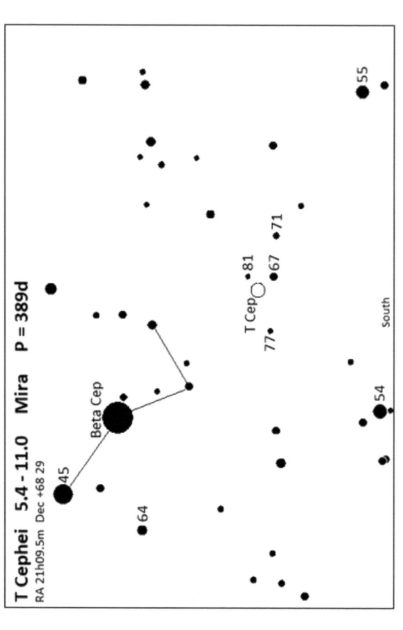

A finder chart, provided by the author, for the Mira-type variable T Cephei and showing magnitudes (with decimal points omitted) for the brighter comparison stars. (Tracie Heywood)

Light Curve for T CEP

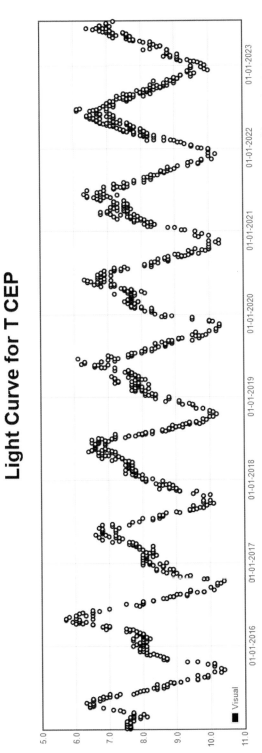

A BAA VSS light curve showing visual observations of the Mira type variable T Cephei from January 2015 to July 2023. (BAA Variable Star Section)

Some Interesting Double Stars

Brian Jones

The accompanying table describes the visual appearances of a selection of double stars. These may be optical doubles (which consist of two stars which happen to lie more or less in the same line of sight as seen from Earth and which therefore only appear to lie close to each other) or binary systems (which are made up of two stars which are gravitationally linked and which orbit their common centre of mass).

Other than the location on the celestial sphere and the magnitudes of the individual components, the list gives two other values for each of the double stars listed – the angular separation and position angle (PA). Further details of what these terms mean can be found in the article *Double and Multiple Stars* published in the 2018 edition of the Yearbook of Astronomy.

Double-star observing can be a very rewarding process, and even a small telescope will show most, if not all, the best doubles in the sky. You can enjoy looking at double stars simply for their beauty, such as Albireo (β Cygni) or Almach (γ Andromedae), although there is a challenge to be had in splitting very difficult (close) double stars, such as the demanding Sirius (α Canis Majoris) or the individual pairs forming the Epsilon (ε) Lyrae 'Double-Double' star system.

The accompanying list is a compilation of some of the prettiest double (and multiple) stars scattered across both the Northern and Southern heavens. Once you have managed to track these down, many others are out there awaiting your attention …

Star	RA		Declination		Magnitudes	Separation	PA	Comments
	h	m	°	'		(arcsec)	°	
Beta[1,2] (β[1,2]) Tucanae	00	31.5	−62	58	4.36 / 4.53	27.1	169	Both stars again double, but difficult
Achird (η Cassiopeiae)	00	49.1	+57	49	3.44 / 7.51	13.4	324	Easy double
Mesarthim (γ Arietis)	01	53.5	+19	18	4.58 / 4.64	7.6	1	Easy pair of white stars
Almach (γ Andromedae)	02	03.9	+42	20	2.26 / 4.84	9.6	63	Yellow and blue-green components
32 Eridani	03	54.3	−02	57	4.8 / 6.1	6.9	348	Yellowish and bluish
Alnitak (ζ Orionis)	05	40.7	−01	57	2.0 / 4.3	2.3	167	Difficult, can be resolved in 10cm telescopes
Gamma (γ) Leporis	05	44.5	−22	27	3.59 / 6.28	95.0	350	White and yellow-orange components, easy pair
Sirius (α Canis Majoris)	06	45.1	−16	43	−1.4 / 8.5			Binary, period 50 years, difficult
Castor (α Geminorum)	07	34.5	+31	53	1.93 / 2.97	7.0	55	Binary, 445 years, widening
Gamma (γ) Velorum	08	09.5	−47	20	1.83 / 4.27	41.2	220	Pretty pair in nice field of stars
Upsilon (υ) Carinae	09	47.1	−65	04	3.08 / 6.10	5.03	129	Nice object in small telescopes
Algieba (γ Leonis)	10	20.0	+19	50	2.28 / 3.51	4.6	126	Binary, 510 years, orange-red and yellow
Acrux (α Crucis)	12	26.4	−63	06	1.40 / 1.90	4.0	114	Glorious pair, third star visible in low power
Porrima (γ Virginis)	12	41.5	−01	27	3.56 / 3.65			Binary, 170 years, widening, visible in small telescopes
Cor Caroli (α Canum Venaticorum)	12	56.0	+38	19	2.90 / 5.60	19.6	229	Easy, yellow and bluish
Mizar (ζ Ursae Majoris)	13	24.0	+54	56	2.3 / 4.0	14.4	152	Easy, wide naked-eye pair with Alcor
Alpha (α) Centauri	14	39.6	−60	50	0.0 / 1.2			Binary, beautiful pair of stars
Izar (ε Boötis)	14	45.0	+27	04	2.4 / 5.1	2.9	344	Fine pair of yellow and blue stars
Omega[1,2] (ω[1,2]) Scorpii	16	06.0	−20	41	4.0 / 4.3	14.6	145	Optical pair, easy
Epsilon[1] (ε[1]) Lyrae	18	44.3	+39	40	4.7 / 6.2	2.6	346	The Double-Double, quadruple system with ε[2]
Epsilon[2] (ε[2]) Lyrae	18	44.3	+39	40	5.1 / 5.5	2.3	76	Both individual pairs just visible in 80mm telescopes
Theta[1,2] (θ[1,2]) Serpentis	18	56.2	+04	12	4.6 / 5.0	22.4	104	Easy pair, mag 6.7 yellow star 7 arc minutes from θ[2]

Star	RA		Declination		Magnitudes	Separation	PA	Comments
	h	m	°	'		(arcsec)	°	
Albireo (β Cygni)	19	30.7	+27	58	3.1 / 5.1	34.3	54	Glorious pair, yellow and blue-green
Algedi (α^{1,2} Capricorni)	20	18.0	−12	32	3.7 / 4.3	6.3	292	Optical pair, easy
Gamma (γ) Delphini	20	46.7	+16	07	5.14 / 4.27	9.2	265	Easy, orange and yellow-white
61 Cygni	21	06.9	+38	45	5.20 / 6.05	31.6	152	Binary, 678 years, both orange
Theta (θ) Indi	21	19.9	−53	27	4.6 / 7.2	7.0	275	Fine object for small telescopes
Delta (δ) Tucanae	22	27.3	−64	58	4.49 / 8.7	7.0	281	Beautiful double, white and reddish

Some Interesting Nebulae, Star Clusters and Galaxies

Brian Jones

Object	RA		Declination		Remarks
	h	m	°	'	
47 Tucanae (in Tucana)	00	24.1	−72	05	Fine globular cluster, easy with naked eye
M31 (in Andromeda)	00	40.7	+41	05	Andromeda Galaxy, visible to unaided eye
Small Magellanic Cloud	00	52.6	−72	49	Satellite galaxy of the Milky Way
NGC 362 (in Tucana)	01	03.3	−70	51	Globular cluster, impressive sight in telescopes
M33 (in Triangulum)	01	31.8	+30	28	Triangulum Spiral Galaxy, quite faint
NGC 869 and NGC 884	02	20.0	+57	08	Sword Handle Double Cluster in Perseus
M34 (in Perseus)	02	42.1	+42	46	Open star cluster near Algol
M45 (in Taurus)	03	47.4	+24	07	Pleiades or Seven Sisters cluster, a fine object
Large Magellanic Cloud	05	23.5	−69	45	Satellite galaxy of the Milky Way
30 Doradus (in Dorado)	05	38.6	−69	06	Star-forming region in Large Magellanic Cloud
M1 (in Taurus)	05	32.3	+22	00	Crab Nebula, near Zeta (ζ) Tauri
M38 (in Auriga)	05	28.6	+35	51	Open star cluster
M42 (in Orion)	05	33.4	−05	24	Orion Nebula
M36 (in Auriga)	05	36.2	+34	08	Open star cluster
M37 (in Auriga)	05	52.3	+32	33	Open star cluster
M35 (in Gemini)	06	06.5	+24	21	Open star cluster near Eta (η) Geminorum
M41 (in Canis Major)	06	46.0	−20	46	Open star cluster to south of Sirius
M44 (in Cancer)	08	38.0	+20	07	Praesepe, visible to naked eye
M81 (in Ursa Major)	09	55.5	+69	04	Bode's Galaxy
M82 (in Ursa Major)	09	55.9	+69	41	Cigar Galaxy or Starburst Galaxy
Carina Nebula (in Carina)	10	45.2	59	52	NGC 3372, large area of bright and dark nebulosity
M104 (in Virgo)	12	40.0	−11	37	Sombrero Hat Galaxy to south of Porrima
Coal Sack (in Crux)	12	50.0	−62	30	Prominent dark nebula, visible to naked eye
NGC 4755 (in Crux)	12	53.6	−60	22	Jewel Box open cluster, magnificent object
Omega (ω) Centauri	13	23.7	−47	03	Splendid globular in Centaurus, easy with naked eye
M51 (in Canes Venatici)	13	29.9	+47	12	Whirlpool Galaxy
M3 (in Canes Venatici)	13	40.6	+28	34	Bright Globular Cluster

Object	RA		Declination		Remarks
	h	*m*	°	‘	
M4 (in Scorpius)	16	21.5	−26	26	Globular cluster, close to Antares
M12 (in Ophiuchus)	16	47.2	−01	57	Globular cluster
M10 (in Ophiuchus)	16	57.1	−04	06	Globular cluster
M13 (in Hercules)	16	40.0	+36	31	Great Globular Cluster, just visible to naked eye
M92 (in Hercules)	17	16.1	+43	11	Globular cluster
M6 (in Scorpius)	17	36.8	−32	11	Open cluster
M7 (in Scorpius)	17	50.6	−34	48	Bright open cluster
M20 (in Sagittarius)	18	02.3	−23	02	Trifid Nebula
M8 (in Sagittarius)	18	03.6	−24	23	Lagoon Nebula, just visible to naked eye
M16 (in Serpens)	18	18.8	−13	49	Eagle Nebula and star cluster
M17 (in Sagittarius)	18	20.2	−16	11	Omega Nebula
M11 (in Scutum)	18	49.0	−06	19	Wild Duck open star cluster
M57 (in Lyra)	18	52.6	+32	59	Ring Nebula, brightest planetary
M27 (in Vulpecula)	19	58.1	+22	37	Dumbbell Nebula
M29 (in Cygnus)	20	23.9	+38	31	Open cluster
M15 (in Pegasus)	21	28.3	+12	10	Bright globular cluster near Epsilon (ε) Pegasi
M39 (in Cygnus)	21	31.6	+48	25	Open cluster, good with low powers
M52 (in Cassiopeia)	23	24.2	+61	35	Open star cluster near 4 Cassiopeiae

M = Messier Catalogue Number   NGC = New General Catalogue Number

The positions in the sky of each of the objects contained in this list are given on the Monthly Star Charts printed elsewhere in this volume.

Astronomical Organizations

American Association of Variable Star Observers

49 Bay State Road, Cambridge, Massachusetts, 02138, USA

aavso.org

The AAVSO is an international non-profit organization of variable star observers whose mission is to enable anyone, anywhere, to participate in scientific discovery through variable star astronomy. We accomplish our mission by carrying out the following activities:

- observation and analysis of variable stars
- collecting and archiving observations for worldwide access
- forging strong collaborations between amateur and professional astronomers
- promoting scientific research, education and public outreach using variable star data

American Astronomical Society

1667 K Street NW, Suite 800, Washington, DC 20006, USA

aas.org

Established in 1899, the American Astronomical Society (AAS) is the major organization of professional astronomers in North America. The mission of the AAS is to enhance and share humanity's scientific understanding of the universe, which it achieves through publishing, meeting organization, education and outreach, and training and professional development.

Association of Lunar and Planetary Observers (ALPO)

Matthew L. Will (Secretary), P.O. Box 13456, Springfield, IL 62791-3456, USA

alpo-astronomy.org

Founded in 1947 by Walter Haas, the ALPO is an international non-profit organization that studies all natural bodies in our solar system. ALPO Sections include Lunar, Solar, Mercury, Venus, Mars, Minor Planets, Jupiter, Saturn, Remote Planets, Comets, Meteors, Meteorites, Eclipses, Exoplanets, Outreach and Online, many with separate "Studies Programs" within these Sections. Minimum membership is very reasonable and includes the quarterly full colour digital *Journal of the ALPO*. Interested observers of any experience are welcome to join. Many members stand ready to improve the skills and abilities of novices.

Astronomical Society of the Pacific

390 Ashton Avenue, San Francisco, CA 94112, USA

astrosociety.org

Formed in 1889, the Astronomical Society of the Pacific (ASP) is a non-profit membership organization which is international in scope. The mission of the ASP is to increase the understanding and appreciation of astronomy through the engagement of our many constituencies to advance science and science literacy. We invite you to explore our site to learn more about us; to check out our resources and education section for the researcher, the educator, and the backyard enthusiast; to get involved by becoming an ASP member; and to consider supporting our work for the benefit of a science literate world!

Astrospeakers.org

astrospeakers.org

A website designed to help astronomical societies and clubs locate astronomy and space lecturers which is also designed to help people find their local astronomical society. It is completely free to register and use and, with over 50 speakers listed, is an excellent place to find lecturers for your astronomical society meetings and events. Speakers and astronomical societies are encouraged to use the online registration to be added to the lists.

British Astronomical Association

Burlington House, Piccadilly, London, W1J 0DU, England

britastro.org

The British Astronomical Association is the UK's leading society for amateur astronomers catering for beginners to the most advanced observers who produce scientifically useful observations. Our Observing Sections provide encouragement and advice about observing. We hold meetings around the country and publish a bi-monthly *Journal* plus an annual *Handbook*. For more details, including how to join the BAA or to contact us, please visit our website.

British Interplanetary Society

Arthur C. Clarke House, 27/29 South Lambeth Road, London, SW8 1SZ, England
bis-space.com

The British Interplanetary Society is the world's longest-established space advocacy organisation, founded in 1933 by the pioneers of British astronautics. It is the first organisation in the world still in existence to design spaceships. Early members included Sir Arthur C Clarke and Sir Patrick Moore. The Society has created many original concepts, from a 1938 lunar lander and space suit designs, to geostationary orbits, space stations and the first engineering study of a starship, Project Daedalus. Today the BIS has a worldwide membership and welcomes all with an interest in Space, including enthusiasts, students, academics and professionals.

Canadian Astronomical Society
Société Canadienne D'astronomie (CASCA)

100 Viaduct Avenue West, Victoria, British Columbia, V9E 1J3, Canada
casca.ca

CASCA is the national organization of professional astronomers in Canada. It seeks to promote and advance knowledge of the universe through research and education. Founded in 1979, members include university professors, observatory scientists, postdoctoral fellows, graduate students, instrumentalists, and public outreach specialists.

Royal Astronomical Society of Canada

203-4920 Dundas St W, Etobicoke, Toronto, ON M9A 1B7, Canada
rasc.ca

Bringing together over 5,000 enthusiastic amateurs, educators and professionals RASC is a national, non-profit, charitable organization devoted to the advancement of astronomy and related sciences and is Canada's leading astronomy organization. Membership is open to everyone with an interest in astronomy. You may join through any one of our 29 RASC centres, located across Canada and all of which offer local programs. The majority of our events are free and open to the public.

Federation of Astronomical Societies

The Secretary, 147 Queen Street, Swinton, Mexborough, S64 8NG
fedastro.org.uk

The Federation of Astronomical Societies (FAS) is an umbrella group for astronomical societies in the UK. It promotes cooperation, knowledge and information sharing and encourages best practice. The FAS aims to be a body of societies united in their attempts to help each other find the best ways of working for their common cause of creating a fully successful astronomical society. In this way it endeavours to be a true federation, rather than some remote central organization disseminating information only from its own limited experience. The FAS also provides a competitive Public Liability Insurance scheme for its members.

International Dark-Sky Association
darksky.org

The International Dark-Sky Association (IDA) is the recognized authority on light pollution and the leading organization combating light pollution worldwide. The IDA works to protect the night skies for present and future generations, our public outreach efforts providing solutions, quality education and programs that inform audiences across the United States of America and throughout the world. At the local level, our mission is furthered through the work of our U.S. and international chapters representing five continents.

The goals of the IDA are:

- Advocate for the protection of the night sky
- Educate the public and policymakers about night sky conservation
- Promote environmentally responsible outdoor lighting
- Empower the public with the tools and resources to help bring back the night

The Planetary Society
60 South Los Robles Avenue, Pasadena, CA 91101, USA
planetary.org

The Planetary Society was founded by Carl Sagan, Louis Friedman and Bruce Murray in 1980 in direct response to the enormous public interest in space, and with a mission to introduce people to the wonders of the cosmos. With a global membership in excess of 50,000 from over 100 countries, it is the largest and most influential non-profit space organization in the world. The Planetary Society bridges the gap between the scientific community and the general public, inspiring and educating people from all walks of life and empowering the world's citizens to advance space science and exploration.

Royal Astronomical Society of New Zealand

PO Box 3181, Wellington, New Zealand

rasnz.org.nz

Founded in 1920, the object of The Royal Astronomical Society of New Zealand is the promotion and extension of knowledge of astronomy and related branches of science. It encourages interest in astronomy and is an association of observers and others for mutual help and advancement of science. Membership is open to all interested in astronomy. The RASNZ has about 180 individual members including both professional and amateur astronomers and many of the astronomical research and observing programmes carried out in New Zealand involve collaboration between the two. In addition the society has a number of groups or sections which cater for people who have interests in particular areas of astronomy.

Astronomical Society of Southern Africa

Astronomical Society of Southern Africa, c/o SAAO, PO Box 9, Observatory, 7935, South Africa

assa.saao.ac.za

Formed in 1922, The Astronomical Society of Southern Africa comprises both amateur and professional astronomers. Membership is open to all interested persons. Regional Centres host regular meetings and conduct public outreach events, whilst national Sections coordinate special interest groups and observing programmes. The Society administers two Scholarships, and hosts occasional Symposia where papers are presented. For more details, or to contact us, please visit our website.

Royal Astronomical Society

Burlington House, Piccadilly, London, W1J 0BQ, England

ras.org.uk

The Royal Astronomical Society, with around 4,000 members, is the leading UK body representing astronomy, space science and geophysics, with a membership including professional researchers, advanced amateur astronomers, historians of science, teachers, science writers, public engagement specialists and others.

Society for the History of Astronomy

Birmingham and Midland Institute, 9 Margaret Street, Birmingham, B3 3BU

shastro.org.uk

The Society for the History of Astronomy was founded in 2002 to promote the study of the history of astronomy by hosting talks by members and publishing new research into the field. One of the main objectives was to encourage research into past astronomers who have previously been neglected within the history of science. Some of its members are professional historians of science but most are amateur historians. The Society hosts several one-day conferences at venues across the United Kingdom each year. A *Bulletin* is published twice yearly containing articles and news items about astronomical history along with short reports of original research by members. The SHA also issues a quarterly electronic newsletter "e-News" which supplements the email messages from the society with updated events/meetings, and general news from council and SHA library. A library of publications of importance to the history of the science is maintained by the Society at the Birmingham and Midland Institute.

One of the Society's major activities is organising a Survey of Astronomical History in the form of lists of historical astronomers and observatories in every part of Britain and Ireland. This has been motivated by a desire to promote research into local astronomical activities that may have previously been neglected. The Society publishes annually a refereed journal called *The Antiquarian Astronomer* containing new research into the history of astronomy, particularly articles written by members. Published papers have discussed activities in major observatories, scientific research by individuals of particular note, scientific instrument makers, and the activities of prominent amateurs.

Society for Popular Astronomy

Secretary: Guy Fennimore, 36 Fairway, Keyworth, Nottingham, NG12 5DU
popastro.com
The Society for Popular Astronomy is a national society that aims to present astronomy in a less technical manner. The bi-monthly society magazine *Popular Astronomy* is issued free to all members.

Webb Deep-sky Society

Secretary: Steve Rayner, 11 Four Acres, Weston, Portland, Dorset, DT5 2JG
webbdeepsky.com
Founded in 1967 – and named after Thomas William Webb, author of *Celestial Objects for Common Telescopes* – the Webb Deep-Sky Society is one of the leading international deep sky organisations, and publishes a journal *The Deep-Sky Observer* together with a regular double star *Circular*. The original aim of the society was to update Webb's publications, and this was achieved through a series of eight handbooks. It still publishes material that it believes is relevant to deep sky observing. The society welcomes all levels of observers and has a number of sections dedicated to the observations of Double Stars, Nebulae and Clusters, and Galaxies.

Our Contributors

Martin Braddock is a professional scientist and project leader in the field of drug discovery and development with 34 years' experience of working in academic institutes and large corporate organizations. He holds a BSc in Biochemistry and a PhD in Radiation Biology and is a former Royal Society University Research Fellow at the University of Oxford. He was elected a Fellow of the Royal Society of Biology in 2010, and in 2012 received an Alumnus Achievement Award for distinction in science from the University of Salford. Martin has published over 170 peer-reviewed scientific papers, filed nine patents, and edited two books for the Royal Society of Chemistry. He also serves as a proposal evaluator for multiple international research agencies. Martin holds further qualifications from the University of Central Lancashire and Open University. He is a member of the Mansfield and Sutton Astronomical Society and was elected a Fellow of the Royal Astronomical Society in May 2015. An ambassador for science, technology, engineering and mathematics (STEM), Martin seeks to inspire the next generation of young scientists to aim high and be the best they can be. To find out more about him visit **science4u.co.uk**

Neil Haggath has a degree in astrophysics from Leeds University and has been a Fellow of the Royal Astronomical Society since 1993. A member of Cleveland and Darlington Astronomical Society since 1981, he has served on its committee since 1989. Neil is an avid umbraphile, clocking up six total eclipse expeditions so far to locations as far flung as Australia and Hawai'i. Four of them were successful, the most recent being in Jackson, Wyoming on 21 August 2017. In 2012, he may have set a somewhat unenviable record among British astronomers - for the greatest distance travelled (6,000 miles to Thailand) to NOT see the transit of Venus. He saw nothing on the day ... and got very wet!

David M. Harland gained his BSc in astronomy in 1977 and his PhD in computer science in 1981. He has lectured in computer science, worked in industry and managed academic research. In 1995 he 'retired' in order to write on space themes.

David Harper, FRAS has had a varied career which includes teaching mathematics, astronomy and computing at Queen Mary University of London, astronomical software development at the Royal Greenwich Observatory, bioinformatics support at the Wellcome Trust Sanger Institute, and a research interest in the dynamics of planetary satellites, which began during his PhD at Liverpool University in the 1980s and continues in an occasional collaboration with colleagues in China. He is married to fellow contributor Lynne Marie Stockman.

Tracie Heywood is an amateur astronomer from Leek in Staffordshire and is one of the UK's leading variable star observers, using binoculars to monitor the brightness changes of several hundred variable stars. Tracie currently writes a monthly column about variable stars for *Astronomy Now* magazine. She has previously been the Eclipsing Binary coordinator for the Variable Star Section of the British Astronomical Association and the Director of the Variable Star Section of the Society for Popular Astronomy.

Rod Hine was aged around ten when he was given a copy of *The Boys Book of Space* by Patrick Moore. Already interested in anything to do with science and engineering he devoured the book from cover to cover. The launch of Sputnik I shortly afterwards clinched his interest in physics and space travel. He took physics, chemistry and mathematics at A-level and then studied Natural Sciences at Churchill College, Cambridge. He later switched to Electrical Sciences and subsequently joined Marconi at Chelmsford working on satellite communications in the UK, Middle East and Africa. This led to work in meteorological communications in Nairobi, Kenya and later a teaching post at the Kenya Polytechnic. There he met and married a Yorkshire lass and moved back to the UK in 1976. Since then he has had a variety of jobs in electronics and industrial controls, and until recently was lecturing part-time at the University of Bradford. Rod got fully back into astronomy in around 1992 when his wife bought him an astronomy book, at which time he joined Bradford Astronomical Society. He is currently working part-time at Leeds University providing engineering support for a project to convert redundant satellite dishes into radio telescopes in developing countries.

Brian Jones hails from Bradford in the West Riding of Yorkshire and was a founder member of the Bradford Astronomical Society. He developed a fascination with astronomy at the age of five when he first saw the stars through a pair of binoculars, although he spent the first part of his working life developing a career in mechanical engineering. However, his true passion lay in the stars and his interest in astronomy took him into the realms of writing sky guides for local newspapers, appearing on local radio and television, teaching astronomy and space in schools and, in 1985, leaving engineering to become a full time astronomy and space writer. His books have covered a range of astronomy and space-related topics for both children and adults and his journalistic work includes writing articles and book reviews for several astronomy magazines as well as for many general interest magazines, newspapers and periodicals. His passion for bringing an appreciation of the universe to his readers is reflected in his writing. You can find out more by visiting his blog at **starlight-nights.co.uk** from where you can also access his Facebook group Starlight Nights.

John McCue graduated in astronomy from the University of St Andrews and began teaching. He gained a PhD from Teesside University studying the unusual rotation of Venus. In 1979 he and his colleague John Nichol founded the Cleveland and Darlington Astronomical Society, which then worked in partnership with the local authority to build the Wynyard Planetarium and Observatory in Stockton-on-Tees. John is currently double star advisor for the British Astronomical Association.

Neil Norman, FRAS first became fascinated with the night sky when he was five years of age and saw Patrick Moore on the television for the first time. It was the Sky at Night programme, broadcast in March 1986 and dedicated to the Giotto probe reaching Halley's Comet, which was to ignite his passion for these icy interlopers. As the years passed, he began writing astronomy articles for local news magazines before moving into internet radio where he initially guested on the Astronomyfm show 'Under British Skies', before becoming a co-host for a short time. In 2013 he created Comet Watch, a Facebook group dedicated to comets of the past, present and future. His involvement with Astronomyfm led to the creation of the monthly radio show Comet Watch, which is now in its fourth year. Neil lives in Suffolk with his partner and three children. Perhaps rather fittingly, given Neil's interest in asteroids, he has one named in his honour, this being the main belt asteroid 314650 Neilnorman, discovered in July 2006 by English amateur astronomer Matt Dawson.

Jonathan Powell worked at BBC Radio Wales as their astronomy correspondent and is currently astronomy and space correspondent for *The National* (an online newspaper for Wales) and a columnist at the *South Wales Argus*. He is also a contributor to CAPCOM, an online magazine which promotes astronomy and spaceflight to the general public, in addition to which he has presented on commercial radio at Sunshine FM in Worcester, Brunel FM in Swindon, Bath FM in Bath, and on the astronomy and space dedicated radio station Astro Radio UK. Jonathan has also written three books on astronomy – *Cosmic Debris: What It Is and What We Can Do About It*; *Rare Astronomical Sights and Sounds* (which was selected by *Choice* magazine as an Outstanding Academic Title for 2019); and *From Cave Art to Hubble: A History of Astronomical Record Keeping*.

Katrin Raynor lives in Pontypridd, Glamorgan, South Wales and is a Fellow of the Royal Astronomical Society and Royal Geographical Society. She is a prolific collector of astronomy and space stamps and is a member of the European Astronomical Society and Astro Space Stamp Society. In her spare time she writes articles and interviews for popular astronomy and philatelic magazines, including *Sky at Night* magazine and *The American Philatelist* for the American Philatelic Society, and is Commissioning Features Editor for the Society for Popular Astronomy's magazine, *Popular Astronomy*.

Peter Rea has had a keen interest in lunar and planetary exploration since the early 1960s and frequently lectures on the subject. He helped found the Cleethorpes and District Astronomical Society in 1969. In April of 1972 he was at the Kennedy Space Centre in Florida to see the launch of Apollo 16 to the moon and in October 1997 was at the southern end of Cape Canaveral to see the launch of Cassini to Saturn. He would still like to see a total solar eclipse as the expedition he was on to see the 1973 eclipse in Mali had vehicle trouble and the meteorologists decided he was not going to see the 1999 eclipse from Devon. He lives in Lincolnshire with his wife Anne and has a daughter who resides in Melbourne, Australia.

P. Clay Sherrod is the director and founder (in 1971) of Arkansas Sky Observatories – the oldest entirely privately-funded astronomical and earth science research facility in the USA. He is the author of 36 books in science and nature, as well as two novels, a book of poetry and a popular cookbook. Sherrod is known worldwide for his creative writings and lectures in the sciences and has contributed over five decades of research in astronomy, palaeontology, archaeology, environmental and earth sciences, and in the fields of molecular biology and research into the origin of life. His publications are detailed at **arksky.org/publications** and the Arkansas Sky Observatories can be accessed at **arksky.org**

Lynne Marie Stockman holds degrees in mathematics from Whitman College the University of Washington and the University of London. She has studied astronomy at both undergraduate and postgraduate levels, and is a member of both the Astronomical Society of the Pacific and the Society for the History of Astronomy. A native of North Idaho, Lynne has lived in Britain since 1992. She was an early pioneer of the World Wide Web; with her husband and fellow Yearbook contributor David Harper, she created the web site **obliquity.com** in 1998 to share their interest in astronomy, computing, family history and cats.

John Vetterlein spent the years of the Second World War at Redbridge, on the eastern approaches to London, and considers himself as fortunate to have survived the V1 flying bomb and V2 rocket attacks. In 1952 he joined the Computing Section of the British Astronomical Association, where he worked on cometary ephemerides and related problems. He went on to become a lecturer in astronomy and the history of mathematics at London University and the University of Surrey, Kingston-Upon-Thames, before moving to Wales in order to take up a post with the University College of Wales, Aberystwyth, where he remained from 1983 to 1990. John Vetterlein is a true polymath. An active musician, he is also a prolific writer, his fiction output alone including sixteen novels, six books of short stories, and a large number of poetry collections, as well as several plays for radio. He has also written monographs on a wide range of topics ranging from peat cutting to binary stars. In addition to all this, he holds the notable distinction of being one of the contributors to the very first *Yearbook of Astronomy*, his article 'Double and Binary Stars' featuring in the 1962 edition. His association with Orkney goes back to 1970 when he visited there to seek out possible sites for auroral work. He is currently engaged on upper atmosphere studies from the Island of Rousay, Orkney, where he has lived for over 30 years.

Society for the
History of Astronomy

All are welcome

Members receive the Society's publications: eNews, Bulletin and Antiquarian Astronomer. There are also meetings and visits to places of interest, use of our unique library and access to research grants.

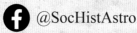 @SocHistAstro

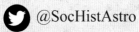 @SocHistAstro

Contact - general.secretary@shastro.org.uk

societyforthehistoryofastronomy.com

Space Oddities Live!

Interested in Astronomy or Space Exploration?
Do you use YouTube or Facebook?
Then check out:

Space Oddities Live!

Streaming Live Every week
Visit our YouTube channel for details!
youtube.com/@spaceoddieslive

Astronomy and Space Exploration news
Discussion
Special Guests
Night Sky Notes
Viewers' Gallery
and more

We livestream every week on YouTube and Facebook. We are an international panel of amateur and professional astronomers. Panel Members are from the UK, Spain, the US and Canada. We chat about anything relating to the Universe and space exploration, keeping our audience up to date with the latest news. We also present interesting presentations on a huge variety of astronomical subjects and also create our own space-related videos for all levels of astronomical knowledge. As well as our weekly shows we also go live for important space launches or other special events. We have a lot of fun, so why not join us? For livestream details, please visit our YouTube Channel at

youtube.com/@spaceoddieslive

For inclusion in our weekly viewers' gallery, please send your images to

spaceoddieslive@gmail.com

Please include your name, location, equipment, processing details etc
One image per email please, entitling it "Gallery Entry"

Background Image Credit: Jonathan Wood and Rachael Wood: Doncaster Astronomical Society

The Commission for Dark Skies

Founded in 1989 by the British Astronomical Association, the Commission for Dark Skies (CfDS) works towards achieving better night skies for all and the protection of the terrestrial environment and biodiversity from the depredations of wasted artificial light.

It is not difficult to define light pollution; put simply, it is light that is wasted, goes to the wrong place and is lighting areas that do not need to be lit. Of course, nobody wants to switch off all lights and live in medieval darkness. The basic solution is to provide the right light, in the right place, at the right time.

Light pollution has increased year-on-year for several decades, and modern bright LED lights are not helping. As well as being too bright for the job many are too white (blue-rich), resulting in damage to the environment and frequent intrusion, creating light nuisance.

The night sky forms half of our visual environment, yet has no protection in law in the UK, unlike despoiling the environment – by fly-tipping for example – which can carry stiff penalties and even land offenders in prison. However, spoiling the beauty of the night sky is often permitted. Light pollution is not just an astronomer's problem. A lot of energy and money is wasted by ill-directed lights. Stray light frays the edges of the habitat of night; contributes to the current biodiversity collapse (most invertebrate species are nocturnal); accelerates climate change; and harms health by disrupting our ancient circadian rhythms.

Most people do not realise the harm that light pollution causes. Everyone has grown up with lighting, both public and private, that is glary, over-bright and shines where it shouldn't. This situation may seem normal, but it is far from natural.

Please get in touch with CfDS via the contacts page at: **britastro.org/dark-skies**

The CfDS works closely with other like-minded groups including:

UK Dark Skies Partnership **ukdarkskies.org.uk**

Dark Sky International **darksky.org** and the UK-based **darksky.uk**

All-Party Parliamentary Dark Skies Group of MPs **appgdarkskies.co.uk**